Machine Learning for Asset Management and Pricing

Machine Learning for Asset Management and Pricing

Henry Schellhorn
Claremont Graduate University
Claremont, California

Tianmin Kong
Claremont Graduate University
Claremont, California

Society for Industrial and Applied Mathematics
Philadelphia

10 9 8 7 6 5 4 3 2 1

Publications Director Kivmars H. Bowling
Executive Editor Elizabeth Greenspan
Acquisitions Editor Elizabeth Greenspan
Developmental Editor Rose Kolassiba
Managing Editor Kelly Thomas
Production Editor Louis Primus
Copy Editor Susan Rodriguez
Production Manager Rachel Ginder
Production Coordinator Cally A. Shrader
Graphic Designer Doug Smock

Library of Congress Cataloging-in-Publication Data
Names: Schellhorn, Henry, author. | Kong, Tianmin, author.
Title: Machine learning for asset management and pricing / Henry Schellhorn, Claremont Graduate University, Claremont, California, Tianmin Kong, Claremont Graduate University, Claremont, California.
Description: Philadelphia : Society for Industrial and Applied Mathematics, [2024] | Includes bibliographical references and index. | Summary: "Covers various machine-learning methods for asset management and asset pricing and attempts to give a unified treatment from the ground up to the latest advances in the field"-- Provided by publisher.
Identifiers: LCCN 2023055176 (print) | LCCN 2023055177 (ebook) | ISBN 9781611977899 (paperback) | ISBN 9781611977905 (ebook)
Subjects: LCSH: Finance--Data processing. | Finance--Mathematical models. | Machine learning.
Classification: LCC HG104 .S34 2024 (print) | LCC HG104 (ebook) | DDC 332.0285--dc23/eng/20231207
LC record available at *https://lccn.loc.gov/2023055176*
LC ebook record available at *https://lccn.loc.gov/2023055177*

To the memory of Jean-Pierre Schellhorn

Henry Schellhorn

To my parents and my cherished wife, Jingchen

Tianmin Kong

Contents

List of Figures

List of Tables

List of Algorithms

Preface

This textbook covers various machine learning methods applied to asset and liability management, as well as asset pricing. We shortened the title to *Machine Learning for Asset Management and Pricing* for practical reasons, but also more fundamental ones. First, we do not give much space to liabilities in this book. It would not render justice to the field of asset and liability management (ALM) to include liabilities in the title. It is, however, important for a student to realize that the comprehensive problem of ALM can be handled (at least in theory) using the same theories and methods as asset management or liability management.

Why not then make a systematic treatise on machine learning to ALM? We think this book is long enough for a semester-long course centered on asset management and pricing. We found out that such a course has its place in the curriculum of graduate students, whether in financial engineering, finance, or quantitative MBA. (The textbook is self-contained and can be used for MS students in other quantitative disciplines, provided they have an interest in finance.) Also, for historical reasons, many financial companies are indeed organized under asset and liabilities lines, and many practitioners end up working only on one side of the divide. Departments that bridge the divide tend to use industry-specific methods. Thus, we refer the reader interested in the big picture to textbooks on risk management for the banking or insurance industry.

This book attempts to subsume the differences between asset management and pricing. The closeness of these two areas is evident. The traditional capital asset pricing model can be used to find *alpha*: an asset with a positive alpha should be bought, whereas an asset with a negative alpha should be sold. This book emphasizes that the same closeness between pricing and management holds for more sophisticated models: a stochastic discount factor that optimally prices assets corresponds to a portfolio return with the highest Sharpe ratio. A cutting-edge model to better estimate prices (or quantify mispricing) should indeed also be a cutting-edge model to manage a portfolio.

Machine learning methods have been deployed in asset management to respond to a major issue facing the classical mean-variance approach. While the optimization techniques that have been used in this model, on both the static side (mean-variance) and dynamic side (strategic asset allocation), are well understood, mean-variance portfolio analysis has yielded disappointing results in practice. One explanation for this phenomenon is behavioral. Strategic asset allocation models tend to produce countercyclical portfolios. It takes a manager a lot of courage to bet against all her peers. Another explanation is statistical: estimation of the mean and of the inverse of the variance-covariance matrix is notoriously unstable. The last twenty years have thus shown a shift of emphasis in research from optimization to econometrics.

Econometrics is classically divided into studying the cross-section of a set of random variables and the time series properties of (usually, but not necessarily) a single stochastic process. This book puts a stronger emphasis on the former than on the latter. Financial econometrics and financial engineering have long focused on studying the time series of (usually a small quantity of) securities, commodities, and other investments. For practitioners, the main purpose was and still is derivatives pricing and hedging, as well as risk management with the use of derivatives. Cross-sectional properties of asset prices are in our opinion harder to comprehend than time series properties. This stems from an obvious advantage of time series data over cross-sectional data: in time series, there is a natural ordering of security prices, namely, by time. There is no such obvious order for a cross-section. The facts that (i) there is only one way to graph a time series and (ii) a human can quickly perceive patterns in a graph explain in a not trivial part the popularity of time series methods and technical analysis. The difficulty for a human to understand a large number of investments is exemplified by the popularity of pair trading. Comparing two investments is easy; more than two is difficult. With the advent of big data, new techniques needed to be developed that go beyond the human ability to comprehend a single graph and make a model with few parameters to explain the data. This is where machine learning comes in. A difficulty in big data is computational complexity. In our experience as academics and quants, the problem is perhaps less to arrive at the result quickly (typical finance problems are comparatively smaller than many other applications of machine learning, and do not need instant decisions) than to find results that are not spurious.

How good are machine learning methods? The signal-to-noise ratio in asset pricing is extremely low compared to other fields, such as image, text, or voice recognition, where machine learning approaches provide amazing results. The signal-to-noise ratio is indeed so low that large numbers of econometricians tried to establish it was negligible, just validating the market efficiency hypothesis. (This fact does not prevent an even larger number of asset managers from trying to "beat the market.") We summarize in one chapter some of the latest evidence of the quality of the models and techniques presented in this book. The papers we reviewed claim that they can improve the Sharpe ratio of a large portfolio by 20%, and sometimes much more.

We had to make some delicate choices. We chose not to present Ross's arbitrage pricing theory [4] since its results can be recovered by other means. However, we provide a rigorous treatment of Lagrange duality. We give two applications of strong duality. The first one is proof of the existence of a risk-neutral measure in the absence of arbitrage. While simpler proofs exist, that proof can be generalized (in some cases) to nonlinear pricing. Also, many advanced results in portfolio theory are based on duality. The second application is to generative adversarial networks, which can sometimes be viewed as solving a dual problem. This book also introduces machine learning methods that are more peripheral to asset pricing. Clustering and principal component analysis have been used in other areas of finance, and students need to know them. Decidedly, not all of finance is trying to beat the market. Finally, we left aside natural language processing as applied to sentiment analysis. We believe it is a field of its own, which deserves another book.

This book originates from class notes of a class that the first author taught twice at Claremont Graduate University as a one-semester course. Most students were graduate students in computer science, financial engineering, economics, or mathematics. Given the variety of students and interests, we kept the level of mathematics and statistics to a minimum, yet every theoretical result is rigorously proved. While all the students had a solid background in multivariate calculus, linear algebra, and probability, many had not been exposed to finance, stochastic processes, statistics, or stochastic calculus. No knowledge of the latter is necessary to read this book. Therefore, we chose to give only a small introduction to option pricing, for which a continuous-time treatment is *de rigueur*, and on which the best textbooks, in our opinion, have already been written.

This book contains more than 120 exercises. Instructors should contact SIAM (textbooks@siam.org) for a solution manual. We tried for most chapters to include at least one computer application in MATLAB or (more rarely) in Excel, or both. When the first author taught this class, students who wanted to code in Python were asked to convert the MATLAB code to Python. The resulting Python code is not included in this book, as we believe converting code is essential for junior programmers to understand the theory better. We built neural networks from scratch without resorting to libraries such as TensorFlow or PyTorch. The reason is as follows. When the first author asked one of his Ph.D. students if he was excited to start an industrial career in machine learning, he answered, "I will be turning some knobs." This textbook is about understanding what is behind the knobs. The knobs will change over time, but the theory and algorithms will remain.

Acknowledgments

We extend our sincere gratitude to the individuals whose invaluable insights and expertise enriched the journey of crafting this textbook. Their steadfast dedication and commitment to excellence have greatly enhanced the content and accuracy of the presented material. We are honored to have collaborated with these contributors and extend our heartfelt thanks.

Specifically, we would like to express our deepest gratitude to

- Mikhail Belkin
- Théodore Economou
- Markus Kalina
- Qidi Peng
- Allon Percus
- Markus Pelger
- Wesley Phoa
- Dominique Salamin
- Rosefinch Fund, Shanghai

Additionally, we acknowledge the broader academic community and our students, whose curiosity and quest for knowledge served as a constant source of inspiration.

Moreover, we thank the numerous individuals who generously contributed their time, expertise, and feedback, helping enhance the quality of this work. Your collective efforts have been instrumental in shaping this textbook into a valuable resource for learners.

Lastly, we would like to express our gratitude to our families and friends for their endless encouragement and patience throughout this endeavor.

We hope this textbook serves as a useful guide and resource in machine learning in asset management and pricing, and inspires a love for learning among its readers.

All errors are ours.

Notation

Scalar variables are usually denoted by lowercase characters a, b, c, but not always. We had to abide by the notation in finance that r usually represents the net return and R the gross return. Vectors are in bold: $\mathbf{x}, \mathbf{y}$. For instance,

$$\mathbf{x} = \begin{bmatrix} x_1 \\ x_2 \\ \vdots \\ x_N \end{bmatrix} \tag{1}$$

Observe that the ith component of a vector $\mathbf{x}$, namely, x_i, is not in bold. Matrices are denoted by uppercase bold characters:

$$\mathbf{A} = \begin{bmatrix} A_{1,1} & \cdots & A_{1,N} \\ \vdots & & \vdots \\ A_{M,1} & \cdots & A_{M,N} \end{bmatrix} \tag{2}$$

Components are denoted by $A_{m,n}$. Since matrices are in bold capital letters, we could not use bold capital letters to denote vector random variables. Thus scalar random variables are in lower case. To distinguish between random variables and their realization we sometimes use the tilde notation for random variables. For instance $P(\tilde{\mathbf{x}} = \mathbf{x})$ is the probability that the random variable $\tilde{\mathbf{x}}$ is equal to $\mathbf{x}$. We will use small letters for indices and the same letter but capitalized for its "logical" upper bound. For instance, (most of the time) i will run from 1 to I, and j from 1 to J. The gradient of a real-valued function $\mathbb{R}^n \to \mathbb{R}$ is the n vector of first partial derivatives and is denoted by $\frac{df}{d\mathbf{x}}$or ∇f. The Hessian matrix is denoted by $\frac{d^2 f}{d\mathbf{x}^2}$. The Jacobian of a vector-valued function $\mathbf{g}(\mathbf{x})$ is the matrix of its first partial derivatives and is denoted by $\frac{d\mathbf{g}}{d\mathbf{x}}$. We use $'$ or ^{t} to denote the transposition of a vector or a matrix. Unless stated otherwise, vectors are column vectors. All our stochastic processes are adapted, that is, known at time t. Rather than writing a recursive relation for the stochastic process x_t in the form

$$x_{t+1} = f(x_t) + u_{t+1}$$

we often write

$$x_{t+1}(\omega) = f(x_t) + u_{t+1}(\omega)$$

as a reminder for the reader that x_t is known but not $x_{t+1}(\omega)$.

The symbols E_t, Var_t, Cov_t denote conditional expectation, variance, and covariance at time t. For one-period models, the time index t is often omitted and will be clear from the context. Sets are in calligraphic font, e.g., $\mathcal{U}$. The cardinality of the set $\mathcal{U}$ is written $|\mathcal{U}|$. We denote by $\mathbf{1}$ the vector of ones.

Applied mathematics abounds with uninteresting exceptions to theorems. Rather than cluttering the text with the assumptions under which a **theorem** is valid, we often choose to call it a **fact**.

Index of Frequently Used Notation

Latin

c_t	consumption at time t
$d_{i,t+1}$	dividend of security i at time $t+1$
f	generic twice differentiable function
$f_{k,t+1}$	kth factor in a K factor model
f_x	probability density function of the random variable x
I	number of securities, (usually) dimension of a datapoint
K	number of parameters
L	Lagrangian function
m_t	stochastic discount factor (SDF)
$M(t)$	stochastic discount factor for excess returns
$n_{i,t}$	number of security i invested in portfolio at time t
$p_{i,t}$	price of security i at time t
r_f	net risk-free rate; equals $R_f - 1$
R_f	(gross) risk-free rate
$r_{i,t+1}$	net return of security i for period t to $t+1$; equals $R_{i,t+1} - 1$
$\bar{r}_i$	expected return of security i
$r_i(t, t+\delta t)$	net return of security i for period t to $t+\delta t$
$r^e_{i,t+1}$	excess return of security i; equals $r_{i,t+1} - r_f$
$r^p_{t+1}(\omega)$	portfolio return
$R_{i,t+1}$	gross return of security i at time $t+1$; equals $\frac{x_{i,t+1}}{p_{i,t}}$
SR_i	Sharpe ratio of security i
T	(time) horizon, number of datapoints in a dataset
u	utility function
u_t	(population) error term at time t
$\hat{u}_t$	sample error term at time t
V	indirect utility function OR cost function in dynamic programming
$w_{i,t}$	proportion of wealth invested in (a.k.a. weight of) security i at time t
$\mathbf{w}^M$	market portfolio
$\mathbf{w}^T$	tangent portfolio
W_t	wealth (of particular portfolio)
$x_{i,t+1}$	payoff of security i at time $t+1$; for stocks equals $d_{i,t+1} + p_{i,t+1}$
$\mathbf{X}$	design matrix, i.e., matrix of values of explanatory variables

Greek

α_i	alpha of security i
β	subjective discount factor
β_i^M	beta of security i with respect to the market
$\beta_{i,k}$	beta of security i with respect to factor $f_{k,t+1}$
μ	portfolio expected return in the mean-variance model

ω	outcome in sample space
$\tilde{\omega}(t, i)$	weight of security i at time t for SDF for excess returns, abbreviated as SDF weight
σ^2	portfolio variance OR variance of the error in homoskedastic models
$\mathbf{\Sigma}_t$	conditional variance-covariance matrix of returns $\mathbf{r}_{t+1}$
$\boldsymbol{\theta}$	parameter vector
$\boldsymbol{\theta}^*$	optimal parameter vector

Chapter 1

Introduction

This chapter starts with a simple example. It then explains the main concepts. We then go into the details of the structure of this book. The last section advises on how to teach from this book.

1.1 ▪ An Example of Asset Pricing: Valuation of a New Firm[1]

Suppose a new newspaper, called the *Claremont Herald*, will open up in Claremont. Investors want to determine how much the *Claremont Herald* is worth as a company. The traditional finance approach is the **net present value** approach: each year the *Claremont Herald* will make a profit, and future profits will be discounted at a particular discount rate (or interest rate). Let r_f denote the risk-free (discount) rate. The net present value of the *Claremont Herald* is

$$NPV = \frac{\text{profit year 1}}{1+r_f} + \frac{\text{profit year 2}}{(1+r_f)^2} + \frac{\text{profit year 3}}{(1+r_f)^3} + \cdots \tag{1.1}$$

The problem of the **fundamental analyst** is then,[2] what profit will the *Claremont Herald* make every year? Clearly, there are two components to the profit:

$$\text{profit} = \text{revenue} - \text{cost} \tag{1.2}$$

Determining the cost is often easier than determining the revenue.[3] Figuring out the revenue is a marketing problem:

- What should be the price of the newspaper?
- What will be the total demand for the newspaper?

A lot of natural intelligence has been spent trying to answer these questions. Here, we will see the **artificial intelligence** or **machine learning** approach.

At the opposite end of the spectrum from the fundamental analyst, or *value investor*, the **technical analyst** (or chartist) will try to determine the future value of the stock of a company by looking at the time series of prices (and trading volume) and identifying some patterns. A typical pattern is head and shoulders (see Figure 1.1). For instance, if such a figure is identified for a stock, then the chartist will conclude that the price of the stock will rise.

[1]This section could also be called "Valuation of a Mispriced Firm."

[2]In this book, we will not model the denominator, i.e., we suppose that the risk-free interest rate r_f is constant.

[3]Together with an accountant or an operations research analyst, the financial analyst can determine today's cost of producing one newspaper and extrapolate this cost to subsequent years.

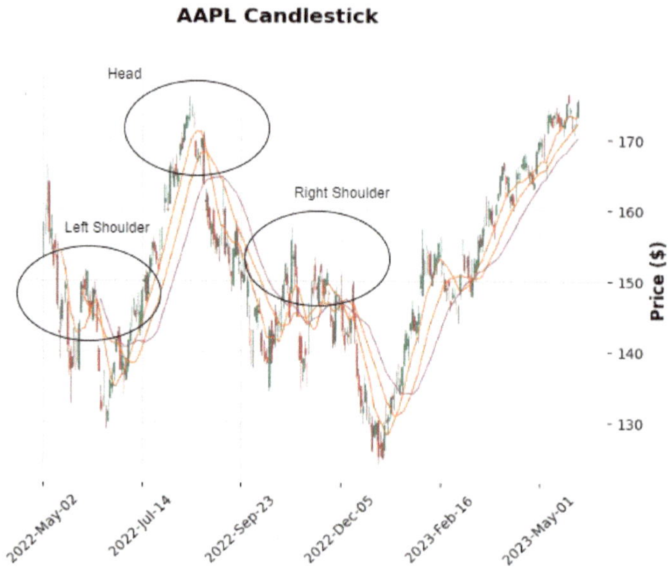

Figure 1.1: The chartist approach: identifying patterns. The figure, which shows Apple stock from 05/22/2022 to 05/21/2023, is an example of a Head and Shoulder pattern.

What does modern finance have to say about fundamental and technical analysis? The **efficient market hypothesis** rebukes both of them, claiming that the price of every asset is right. Otherwise, there would be arbitrage: market participants would sell assets that are too dear and buy assets that are too cheap, thus bringing back the price of each security to its **fundamental value**. The efficient market hypothesis has been shown empirically to be correct for stocks of large companies in stable sectors.

Yet, thousands of analysts work in fundamental analysis, technical analysis, or **statistical arbitrage**. What is statistical arbitrage? A typical example is **pair trading**. Going back to our example, an investor (call him Bill) finds the business plan of the *Claremont Herald* interesting but may be reluctant to invest in a newspaper, considering that newspapers are risky businesses. We say there is a **systematic** risk to investing in the newspaper business. Bill will identify a *similar* newspaper, say the *Purdue Bee*, which he believes has weaker prospects. The trading strategy will thus be

$$\text{buy } \textit{Claremont Herald} \text{ / sell } \textit{Purdue Bee} \tag{1.3}$$

This pair trading strategy would thus eliminate the systematic newspaper risk. These investment strategies can be effective, yet they do not quite answer the question we had at the beginning: what should be the value of the *Claremont Herald*?

As the *Claremont Herald* is not traded yet, we can dismiss both the efficient market hypothesis and technical analysis to find its price.[4] In this case, one avenue is to **price by arbitrage**, which is a variant of statistical arbitrage. The steps are as follows:

1. Identify what similar firms have the right stock price.

2. Identify what **features** of these firms are relevant for pricing.

[4]More generally, one should not pay too much attention to the efficient market hypothesis or technical analysis when securities are very thinly traded.

3. Use the share prices of these firms, as well as their features, to **train** a pricing model.

4. **Apply** the model to price the *Claremont Herald.*

Of course, this arbitrage pricing methodology can also be used to make investment decisions for securities that are traded: you may believe that a particular firm (say *The New York Times*) is not correctly priced and apply your pricing model on it. If *The New York Times* **model price** is more than its current stock price, then the investor's decision should be to buy shares of *The New York Times*.

A Digression on Machine Learning

While pair trading is a type of strategy that can easily be done by a human investor, the universe of all arbitrage strategies is best explored by **machine learning** (ML). Supervised learning is an area of ML where training data is composed of

- features
- labels.

For instance, in image processing, the features would be images of animals, and the label would be their species. In the example below (see Figure 1.2), the images of a fish or a horse are used to train the model. We then apply the model to identify a hippocampus. The model may or may not correctly identify it as a fish. If it does not, we say that the model has **generalization error**.

Similarly, supervised learning could be used to price the *Claremont Herald*, following the steps shown on the previous page. For the first step, **clustering** can help in identifying similar companies.

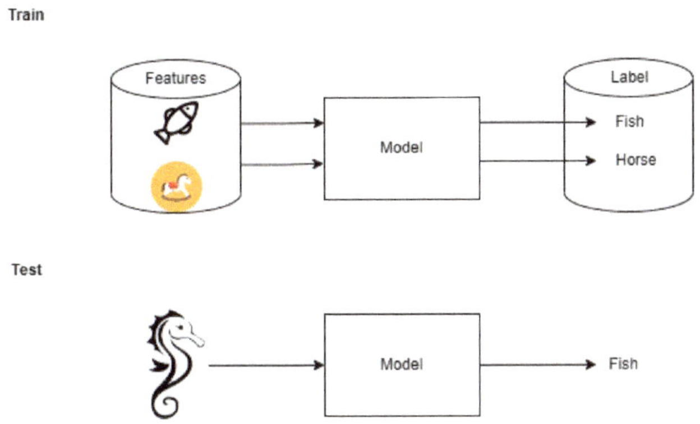

Figure 1.2: Image recognition by machine learning: the hippocampus is classified correctly as a fish.

...And Back to Asset Pricing

For the second step, namely identifying company features for pricing, **modern portfolio theory** can help us. Among other things, modern portfolio theory shows that only **systematic risk** should be priced by the market, while the other risks (called **idiosyncratic risks**) should not. For instance, in the *capital asset pricing model* (**CAPM**) only one feature of a company, namely, the beta of the company (with respect to the market return), matters to price its stock. To describe the CAPM, consider the following **linear regression** of stock excess[5] return on the market excess return:

$$\text{stock excess return} = \text{alpha} + \text{market beta} * \text{market excess return} + \text{noise} \tag{1.4}$$

The CAPM asserts that in equilibrium

- **alpha** $= 0$
- the noise is uncorrelated with the stock excess return.

Thus, if the alpha is positive, the security is mispriced: it will make superior returns on average and should be bought. Portfolio management is often called **a quest for alpha**. In this particular model, the market excess return[6] is called a **factor**. The **arbitrage pricing theory** recognizes that more than one factor can explain the returns of securities.[7] A **K-linear factor model** is of the form

$$\text{stock } i\text{'s excess return}(t+1) = \alpha_i + \sum_{k=1}^{K} \beta_{i,k} f_k(t) + \text{ noise}(t+1) \tag{1.5}$$

In other words, the excess return of a security is explained by its features (its alpha α_i and its beta with respect to factors, namely, $\beta_{i,k}$).

Let us go back to the problem of pricing the *Claremont Herald*. How are we going to price it, given that we found the alphas and betas of *The New York Times* and *Purdue Bee*? One approach would be ad hoc interpolation. Another one would be a structural model of β for a representative firm. Yet, a more powerful methodology is at hand.

Beyond Linear Factor Models: The Stochastic Discount Factor (SDF) Methodology

When there is no arbitrage, in particular for option pricing, the SDF methodology should be applied. The SDF is calibrated by minimizing the mispricing of all the liquid securities. It is a **positive** stochastic process, denoted by $m(t)$. The **payoff** of a stock is equal to its dividend plus its stock price. We will prove that if there is no arbitrage, the value of a share in any company is the expected value (across all states of the world) of the SDF times the payoff:

$$\text{Value of the company at time } t = E_t[m(t+1) * \text{payoff } (t+1)] \qquad t \in [1, T] \tag{1.6}$$

where E_t is the conditional expectation at time t. We will also prove in this book that there exists a constant γ_t known at time t such that

$$\frac{m(t+1)}{\gamma_t} = 1 - \sum_{j=1}^{\text{Number of companies}} \tilde{\omega}(t,j) * \text{excess return of company } j \text{ in } [t, t+1] \tag{1.7}$$

[5]The excess return is equal to the return minus r_f.

[6]To fix orders of magnitude, the market return in the US is of the order of 10%. Subtracting the risk-free interest rate r_f yields a market excess return of the order of 7 to 8%.

[7]For instance, in the Fama–French three-factor models, the factors are (i) market excess return, (ii) return of high (market cap) minus low (market cap), (iii) return of high (book-to-market ratio) minus low (book-to-market ratio). The values of these three factors are denoted by $f_k(t)$ with $k \in \{1, 2, 3\}$.

While the excess stock return of company i in $[t, t+1]$ is not known at time t, the importance of this formula resides in the fact that the **SDF weight** $\tilde{\omega}(t, j)$ can be determined at time t. Putting these formulas together yields

$$\gamma_t * \text{Value at } t \text{ of firm } i = E_t[\text{payoff firm } i] - \sum_{j=1}^{\text{Number of companies}} \tilde{\omega}(t, j) E_t[\text{payoff} * \text{excess return company } j \text{ at time } t+1] \quad (1.8)$$

An approach to calibrate the SDF weight is to use a representative firm model, that is,[8]

$$\tilde{\omega}(t, j) = \tilde{\omega}(\text{systematic info at } t, \text{ idiosyncratic info of company } j \text{ at } t) \quad (1.9)$$

Thus we will need to estimate **one formula** for the SDF weight, based on a mass of data. Such an SDF approach merges the three most important techniques in pricing:

- fundamental analysis: for idiosyncratic information, one takes accounting variables
- technical analysis: for systematic information, as well as the conditional expectation terms in (1.8), one takes market variables
- no-arbitrage theory.

Where does machine learning come in? In Chen, Pelger, and Zhu (2020) [5], the formula for the SDF weight is a **nonlinear** function of the parameters since it is the output of a **neural network**. Nonlinear function approximation is more general than linear function approximation and should yield better results.

Implications for Asset Management

A key indicator for asset management is the **Sharpe ratio**. Asset managers want to maximize the Sharpe ratio. We will prove that, if we invest the **SDF weight** $\tilde{\omega}(t, j)$ in each security, then the resulting portfolio maximizes the Sharpe ratio. Thus knowing the precise form of the SDF weight (as explained in the example above) as a function of the information available at time t has two important applications:

- pricing
- portfolio management.

Formula (1.8) is probably not applied in many quant shops for pricing. What about portfolio management? Chen, Pelger, and Zhu (2020) [5] report a Sharpe ratio of 0.75 using deep learning of the SDF weight, which is **20% higher than most quantitative funds.**

1.2 ▪ Concepts: An Overview

This book is firmly rooted in **modern portfolio theory** (also termed mean-variance analysis), pioneered by Markowitz in 1952, for which he received the Nobel Prize, as well as to its pricing

[8] If one views a formula as applying *parameters* to *input data* or features, the parameters are firm-independent, but the features can be firm-dependent. Note the abuse of notation between $\omega(\tilde{t}, j)$ and $\tilde{\omega}$ (information on j at t).

counterpart, the CAPM (Sharpe [70]). They are direct applications of convex optimization and linear statistical models. In our opinion, every student should know the classical treatment of these models. The concepts of diversification of investments and the primary role of the covariance between asset returns and factors to explain prices come from these groundbreaking advances. Yet other concepts have emerged.

The main concepts we articulate this book around come from several fields: statistics, optimization,[9] machine learning, finance, and control theory. From statistics, we borrow two main concepts: ergodicity and conditional estimation. When a time series is **ergodic**, time averages converge to the population mean. This property is critical in financial applications, as we cannot repeat experiments, unlike other areas of statistics. **Conditional estimation** often yields results superior to those of unconditional estimation: indeed, there are more ways to improve the estimation of conditional moments than to improve the estimation of unconditional moments. As a result, the coefficients we obtain from either a linear regression or a nonlinear regression can be made a function of contemporaneous information and thus can yield superior performance in asset management. From optimization, we highlight the concept of (Lagrange) **duality**. From machine learning, we retain mainly three concepts: regularization, boosting, and nonlinear function estimation. **Regularization** of a model results in smoother parameters, i.e., parameters that are usually more robust. In the parlance of ML, this means a smaller generalization error. The idea of **boosting** estimators (by focusing on what is important) is used in several algorithms of this book: Cochrane–Orcutt, GMM, and GAN. The greater generality of **nonlinear functions** (of the parameters) over linear ones explains the success of nonlinear function approximation techniques, such as neural networks and random trees and forests. Finance provides us with the concept of **no-arbitrage**, which is related to, but often confused with, the law of one price. Roughly speaking, there are two slightly different methodologies in asset pricing: factor models, which use the law of one price, and stochastic discount factor methods, which use the no-arbitrage condition. In predictive methods, one tries to forecast the return of securities, or more exactly to estimate the expected return. The law of one price tells us that the alpha of security should be zero, and, if it is not, then mispricing occurs, which should be exploited. The assumption of no-arbitrage, i.e., that no superior profit can be gained without taking extra risk, puts additional restrictions on the model, or more precisely on the structure of the stochastic discount factor: it should be positive. Finally, control theory (or its machine learning cousin, reinforcement learning) has been extensively studied in academia since the famous 1973 paper by Nobel Prize winner Merton [1]. Brennan, Schwartz, and Lagnado [2] coined in 1997 the term **strategic asset allocation** to the application of control techniques to the portfolio problem, while academics prefer the term dynamic portfolio theory. This theory is mathematically exciting and sophisticated; however, the complexity in the estimation of the models and the difficulty for practitioners to justify the trading strategy stemming from the model still delay their deployment in the industry. For that reason, this book only gives a cursory treatment of strategic asset allocation. A reader who wants to specialize in the new advances that ML brings to dynamic portfolio theory is referred to the recent book *Machine Learning in Finance* [50].

This book tries to highlight the commonality of the mathematics between the machine learning section and the asset management and pricing section. To give a few examples,

- duality explains at the same time the no-arbitrage condition and the architecture of generalized adversarial networks

- the importance of covariance applies both to the CAPM and to regularization.

[9]For convenience, we classified statistics and optimization under machine learning in the title of the book. It can also be argued that machine learning is only a (computational) extension of these other two fields.

1.3 ▪ Structure of the Book

This book can be read sequentially. However, it is more interesting for many readers to mix theory (machine learning) with application (asset pricing and management). These readers should consult Figure 1.3, which explains the interdependencies between the chapters of the book. The next section provides two course plans that exemplify the strategy to mix and match these different topics.

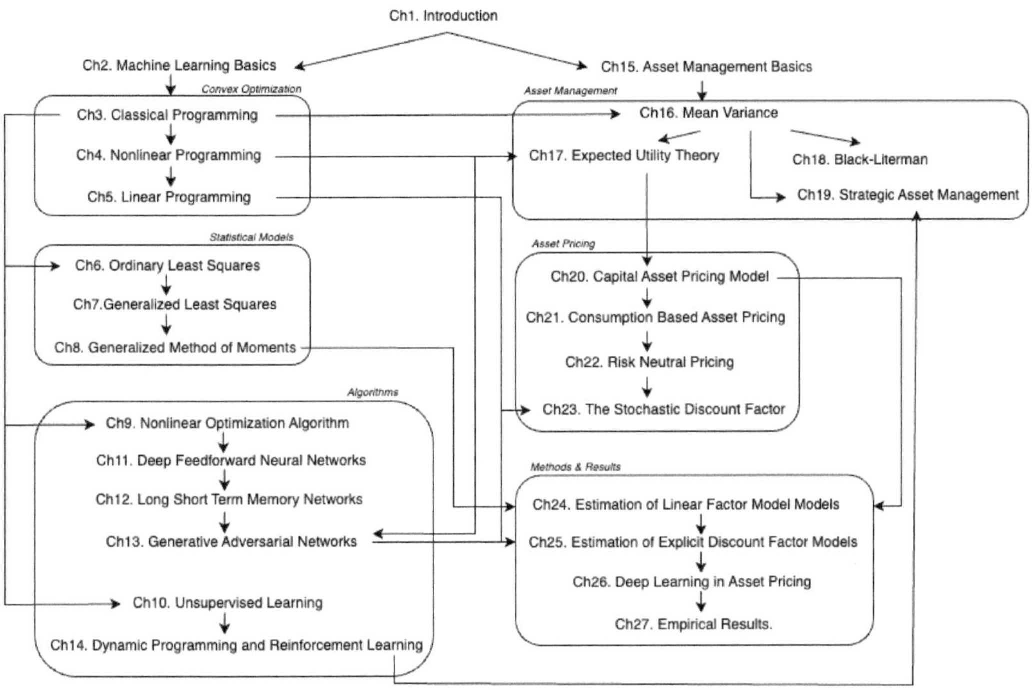

Figure 1.3: How to read this textbook.

The **machine learning** part of the book consists of three main subparts: convex optimization, linear statistical models and methods, and algorithms. The chapters on **convex optimization** are classical programming, nonlinear programming, and linear programming. These topics are usually covered in an *optimization* or *mathematical programming* course. These chapters have a mathematical emphasis that hopefully will not discourage the reader. The chapters on **linear statistical models and methods** are ordinary least squares (OLS), generalized least squares (GLS), and generalized method of moments (GMM). These classical techniques are rooted in convex optimization. While some authors reserve the term "machine learning" for more modern techniques, linear models are still largely dominant in asset pricing and deserve to be studied, with an eye towards generalization to the nonlinear case. The chapters on **algorithms** cover both supervised and unsupervised learning. The emphasis is, however, on supervised learning, and only one chapter is dedicated to unsupervised learning. The latter needs to be read, if only because it introduces the fundamental principle of estimation theory. After a chapter on nonlinear optimization algorithms, this book covers neural networks in three chapters: deep feedforward neural networks, long short-term memory (LSTM) networks, and generative adversarial networks. At the time of writing this book, many students are attracted to LSTM networks because they can easily visualize the output. To us, most of the focus on LSTM can be considered data mining,[10] and we prefer to center the chapter on LSTM networks on classical (linear) models, which are less prone to generalization error.[11] The chapter on dynamic programming and reinforcement learning is more complicated than the others and can be skipped at first reading.

The **asset pricing and management** part of the book is divided into three parts: asset management, asset pricing, and methods and results. We start with **asset management**. As in microeconomics, the individual optimization problem is easier than the general equilibrium and pricing problem and needs to be covered first. The chapter on mean-variance portfolio analysis is the classical Markowitz theory. The chapter on expected utility theory comes next; it considers the full probability distribution of returns and goes beyond the first two moments. The chapter on Black–Litterman introduces a Bayesian technique that is popular with practitioners who need to implement the mean-variance approach. The chapter on strategic asset management considers the dynamic problem. It applies the concepts from dynamic programming.

There are four chapters on **asset pricing**. The CAPM is the natural outcome of mean-variance analysis. Consumption-based asset pricing is a more theoretical chapter, which uses stylized assumptions of the behavior of agents. Its main message is that the SDF should be positive. The risk-neutral pricing chapter applies these findings to the pricing of stocks, bonds, and options. The SDF chapter employs two approaches (that are now dominant among practitioners) to justify risk-neutral pricing: the law of one price, which does not require the SDF to be positive, and the no-arbitrage approach, which does.

The final subdivision on the book on **methods and results** starts with the estimation of linear factor models of return, which are currently dominant in industry, perhaps because of their mathematical simplicity. In the chapter on the estimation of the explicit discount factor, we focus on the generalized method of moments. However, a more direct ordinary least squares approach can also be used. The word "explicit" is meant to remind us that models of return and the SDF are equivalent. The advantage of the latter is that it is easier to impose a positivity constraint on the SDF directly, thus guaranteeing the absence of arbitrage in the model. The chapter on deep learning in asset pricing focuses on neural networks. We conclude with a chapter on empirical results, where we mostly compare the latest results from the literature.

1.4 ▪ Course Plans

When the first author taught this material, the class notes, from which this book originates, were terse. To alleviate this terseness, students were asked to read the first part of the book *Asset Management: A Systematic Approach to Factor Investing* by Andrew Ang [6]. This is a classic MBA text which provides all the financial intuition needed for better appreciating this book. Instructors in other countries than the US may want to resort to different sources.

This book contains a large quantity of exercises. Instructors should contact SIAM (textbooks@siam.org) for a solution manual. Since students and readers have a large variety of programming experience, we decided to label each exercise difficulty (as "easy," "medium," and "hard") only for mathematical exercises. We also drafted some projects, which are supposed to take much more time to complete than exercises. One project in our opinion is so long that it could serve as a basis for a paper or a Ph.D. dissertation.

We present two different course plans, the first one for students with little background in optimization, the other one for more advanced students.

1.4.1 ▪ For Students with Little Background in Optimization

Table 1.1 provides a sample study plan for students with little background in optimization; we omit Chapters 14 and 19, which are more technical.

[10]That has already been mined.

[11]We refer the reader to the book by Lopez De Prado [3] for the latest advances in the field.

Week	Chapters
1	1, 2
2	3, 6
3	4, 16
4	7, 17, 20
5	5, 9
6	21, 22
7	23
8	11, 12
9	8, 13
10	25, 26
11	midterm examination
12	24, 27
13	18
14	10
15	student presentations

Table 1.1: Study plan for students with little background in optimization.

The reason we pushed Chapters 24, 27, 18, and 10 towards the end of the course is that we assigned students projects on neural networks, and we wanted the material covered by week 10, so they can start their project then.

1.4.2 ▪ For Advanced Students

Table 1.2 provides a sample study plan for advanced students. We omit Chapters 3, 4, and 5, which are prerequisites of such a course.

Week	Chapters
1	1, 2
2	6, 16
3	7, 17, 20
4	9, 21
5	22, 23
6	11, 12
7	8, 13
8	25, 26
9	midterm examination
10	24, 27
11	18
12	10
13	14
14	19
15	student presentations

Table 1.2: Study plan for advanced students.

As in the previous course plan, we frontload the neural networks material so that students can start projects on it around week 8.

Part I

Machine Learning

Chapter 2

Machine Learning Basics

At the time of writing this book, *artificial intelligence (AI)* has caught the attention of probably everybody on the planet. Successes have been tremendous in speech recognition, voice recognition, intelligent conversation (with Google Bard and OpenAI GPT4), and many other applications.

AI followed the computer revolution. The initial idea was to build *expert systems*, where computers would search through a database of facts guided by intelligent human programming. While this area remains an important part of AI under the more modern name of *knowledge bases*, machine learning (ML) emerged as a more promising (for some applications) area of AI. In ML, the machine learns by itself, with limited interaction from human programmers.

An ML algorithm is an algorithm that is able to learn from data. Mitchell in [7] provides a complete definition: "A computer program learns from experience E with respect to some class of tasks T and performance measure P if its performance at tasks in T, measured by P, improves with experience E." Let us look now at some tasks that ML can perform, performance measures by which algorithms are judged, and the type of experience these algorithms utilize.

2.1 ▪ Definitions

2.1.1 ▪ The Task T

The following tasks can be handled by ML:[12]

- classification
- classification with missing inputs
- regression
- transcription: for instance, speech recognition
- machine translation
- structured output: this is a generalization of translation and speech recognition, where the output is structured
- anomaly detection: for instance, credit card fraud detection

[12]For more information on the topic, we refer the reader to Chapter 5 in [8], from which much of the material in this section is taken. We, of course, emphasize in this book the techniques that are useful in asset management and pricing.

- synthesis and sampling: for instance, creating scenery for video games
- imputation of missing values (interpolation)
- denoising
- (probability) density estimation.

In the finance industry, regression, anomaly detection, and density estimation are the most common tasks. In this book, we will focus mainly on regression, that is, the estimation of a function $f^* : \mathbb{R}^I \rightarrow \mathbb{R}$. A typical task is to estimate the expected return of a security over the next period based on current and past information. This will be described in Chapter 24. Another closely related task is the determination of the stochastic discount factor, which will be described in Chapters 25 and 26.

Function estimation is a well-known problem in applied mathematics. Most of the field of *approximation theory* is dedicated to it. When data is sampled from a probability distribution, function estimation is one of the central problems of statistics. We will review linear statistical models in Chapters 6, 7, and 8. Most scientists would agree that the classical linear models can be considered to be part of ML. What more recent ML techniques added to linear models is the ability to estimate

- nonlinear functions of the parameters
- on very large problems, i.e., problems with a very large number of parameters and/or data.

2.1.2 ▪ The Experience E

Machine learning algorithms can be either **supervised** or **unsupervised**. All algorithms experience a **dataset** or **sample**. A dataset is a collection of many **examples** or **data points**. An example of $\mathbf{x} \in \mathbb{R}^I$ is a collection of **features** x_i. In this book, we will very often use examples taken at different times and label them by t. Thus the dataset is

$$\{\mathbf{x}_1, \ldots, \mathbf{x}_t, \ldots, \mathbf{x}_T\}$$

Typical tasks in **unsupervised learning** are probability density estimation, clustering, and principal component analysis.

In **supervised learning**, each example $\mathbf{x}$ is associated with a **label** or a **target** $\mathbf{y}$, which can be a scalar. For instance, $\mathbf{x}$ is the list of 150 features of a single flower plant. The goal is to classify a new flower. The ML algorithm is trained with many examples of flowers where y is the label of every example (iris, rose, tulip, ...).

Some ML algorithms do not experience a fixed dataset. **Reinforcement learning** algorithms interact with an environment, so there is a feedback loop between the learning system and its experiences. These are discussed in Chapter 14.

2.1.3 ▪ The Performance Measure P

Different tasks have different performance measures. We thus focus here on the performance of regressions (with scalar target y_t) defined as the problem of estimating a function $f^* \, \mathbb{R}^I \rightarrow \mathbb{R}$ that generates the data:

$$y_t = f^*(\mathbf{x}_t) + u_t$$

where u_t is often called the noise, or intrinsic error, that is, the error that cannot be offset by the clever use of ML. It is *irreducible*. The statistician will determine an approximation $f(\mathbf{x}_t, \boldsymbol{\theta})$ of the target function $f^*(\mathbf{x}_t)$ with the parameter vector $\boldsymbol{\theta} \in \mathbb{R}^K$. Suppose the algorithm returns a suboptimal vector $\widehat{\boldsymbol{\theta}}$ for the dataset to which correspond fitted values:

$$\hat{y}_t = f(\mathbf{x}_t, \widehat{\boldsymbol{\theta}})$$

The sample error $\hat{y}_t - y_t$ is usually measured by the **mean squared error**, which is defined by

$$\text{MSE} = \frac{1}{T}\sum_{t=1}^{T}(\hat{y}_t - y_t)^2$$

Another measure is the sum of squared errors (SSE), which is equal to the MSE times the number of observations. Obviously, it has the same informational content as the MSE. Another measure widely used in linear models is R-squared, or R^2. It is formally defined in Chapter 6.

Performance for Parametric Models

In **frequentist statistics** the true parameter vector $\boldsymbol{\theta}^*$ is an unknown constant, while the dataset is "random." This is the traditional viewpoint, which often puzzles the beginner: why would something we know (the sample values) be random while something we do not know (the parameter we try to estimate) is nonrandom? **Bayesian statistics** has the opposite standpoint. For Bayesians, the parameter vector is random, and a dataset sample is nonrandom. This book uses Bayesian statistics only in Chapters 18 and 24. We will here focus on the frequentist interpretation.

Definitions. A *point estimator* or *statistic* is any function of the data:

$$\hat{\boldsymbol{\theta}}_T = g(\mathbf{x}_1, \ldots, \mathbf{x}_t, \ldots, \mathbf{x}_T)$$

Since the dataset is random, the estimator $\hat{\boldsymbol{\theta}}_T$ is a random variable. As explained in the notation section, we sometimes highlight its randomness by using the notation $\hat{\theta}_T(\omega)$.[13] We can thus define the *bias* of the estimator,

$$\text{Bias}(\hat{\boldsymbol{\theta}}_T) = E[\hat{\boldsymbol{\theta}}_T] - \boldsymbol{\theta}^*$$

and, if the estimator is a scalar, the variance of the estimator,

$$\text{Var}(\hat{\theta}_T) = E[(\hat{\theta}_T - E[\hat{\theta}_T])^2]$$

The population equivalent of the MSE of the estimator, namely, $E[(\hat{\theta}_T - \theta^*)^2]$, can be decomposed into bias and variance, as you shall prove in Exercise 2.1:

$$E[(\hat{\theta}_T - \theta^*)^2] = \text{Bias}(\hat{\theta}_T)^2 + \text{Var}(\hat{\theta}_T) \tag{2.1}$$

While a large proportion of estimators constructed by statisticians are unbiased, some achieve lower MSE by reducing the variance and increasing the bias.

2.1.4 ▪ Capacity, Overfitting, and Underfitting

The main challenge in ML is that the algorithm must perform well on *new, previously unseen* inputs, not just those on which our model was trained. The ability to perform well on previously unobserved input is called **generalization**.

[13]Technically wrong.

For instance, in nonlinear regression the error $y_t - \hat{y}_t$ can be decomposed into

$$\underbrace{y_t - \hat{y}_t}_{\text{error}} = \underbrace{f^*(\mathbf{x}_t) - f(\mathbf{x}_t, \boldsymbol{\theta}^*)}_{\text{approximation error}} + \underbrace{f(\mathbf{x}_t, \boldsymbol{\theta}^*) - f(\mathbf{x}_t, \hat{\boldsymbol{\theta}})}_{\text{estimation error}} + \underbrace{u_t}_{\text{intrinsic error}}$$

Estimation error is determined by the data. The more data, the smaller the estimation error. Approximation error is controlled by the statistician, i.e., by the choice of the class of function $\{f(., \boldsymbol{\theta})\}$ for all possible $\boldsymbol{\theta}$. The more general the class of functions, the smaller the approximation error for a given sample. However, the error is potentially larger on another set.

When training an ML model, we have access to a **training set**. The error on the training set is called the training error. What separates ML from optimization is that we want the **generalization error**, also called **test error**, to be low as well. The test error is the expected value of the error on a new input, called the **test set**. In finance and economics, this corresponds to the traditional difference between **in-sample** and **out-of-sample** performance of a model. Additionally, in ML, a third dataset is needed, namely, the **validation set**, as we will see in another section. The requirements for a good ML algorithm are to

- make the training error small
- make the gap between training and test error small.

Underfitting occurs when the model is not able to obtain a sufficiently low error value on the training set. **Overfitting** occurs when the test error is much larger than the training error.

We can control whether a model is more likely to overfit or underfit by altering its **capacity**. A model's capacity is its ability to fit the training set. Models with a high capacity can overfit by memorizing properties of the training set that do not fit them well on the test set. A linear regression of y on x has less capacity than a regression of y on the basis $\{1, x\}$, that is, on all affine functions. The latter has less capacity than a regression of y on the quadratic functions in x, and so forth. The quality of the fit depends on the relation between the size T of the dataset and the number K of features. For linear models (see Chapter 6),

- if $T < K$ we have overfitting
- if T is slightly larger than K the model has (generically) the appropriate capacity
- if $T \gg K$ we may have underfitting.

Figure 2.1 illustrates these three different cases.

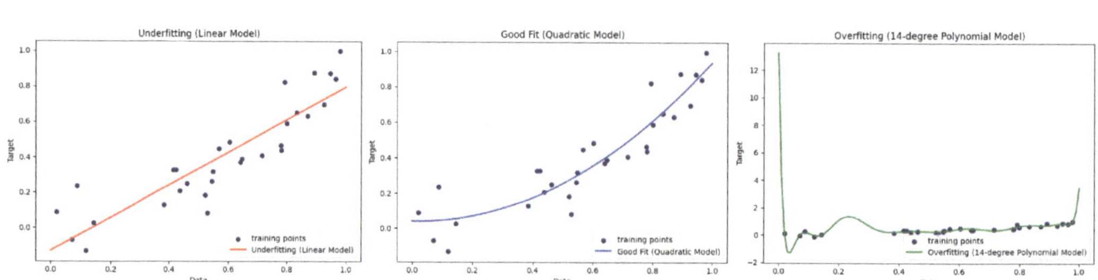

Figure 2.1: Underfitting vs. overfitting. The data is generated from a quadratic function. A linear function underfits the model. A quadratic function (in the middle) has the appropriate capacity, while a degree 9 polynomial (the figure to the right) overfits the data.

The decomposition (2.1) of the MSE of an estimator into bias and variance gives a different light to the tradeoff approximation error/estimation error as a function of the capacity of the model. Most researchers believe that there exists a "sweet spot" between underfitting and overfitting; see Figure 2.2.

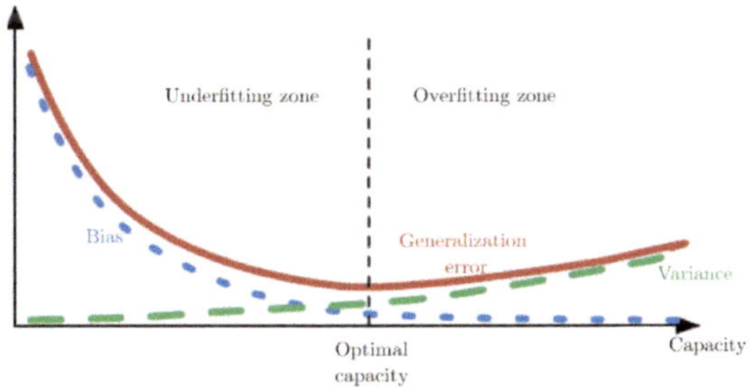

Figure 2.2: The "sweet spot" between underfitting and overfitting. From Figure 5.6 in [8].

Some researchers believe that there are two sweet spots and that overfitting can be good, as illustrated in Figure 2.3. The fact that several successful models seem to be overparameterized gives credence to this intriguing (yet) unconventional hypothesis. Indeed Google's language model has over one trillion parameters [9]. Some researchers refer to it as "neural collapse" [10].

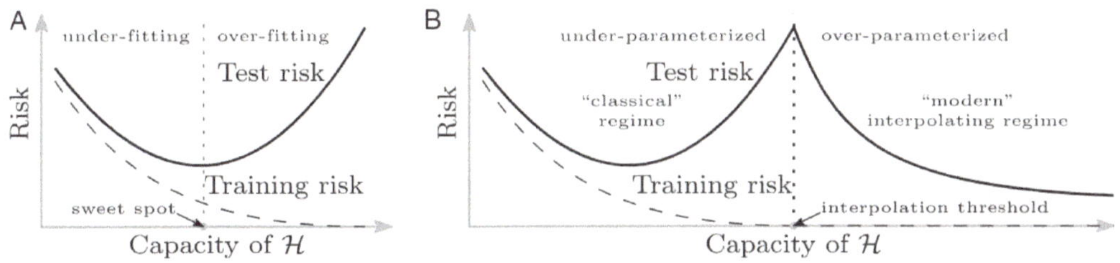

Figure 2.3: Two "sweet spots" between underfitting and overfitting. From Figure 1 in [11]. Reprinted with permission from the National Academy of Sciences. Belkin, M., Hsu, D., Ma, S., and Mandal, S., "Reconciling modern machine-learning practice and the classical bias-variance trade-off," Proceedings of the National Academy of Sciences, Volume 116, Issue 32, 15849–15854, 2019.

How good is ML? This answer is, not good in general. The **no free lunch theorem for machine learning** in [12] states that, averaged over all possible test sets, every classification algorithm has the same test error rate. In other words, no ML algorithm is universally better than any other. However, if we make assumptions about the kinds of test sets that we encounter in real-world applications, then we can design ML algorithms that perform well on these test sets.

2.2 ▪ Regularization

One way to (softly) restrain the capacity of our model is to add a **regularizer**. If we want to minimize the number of components of θ that are nonzero we would use **LASSO**, where the regularizer takes the form

$$REGULARIZER(\boldsymbol{\theta}) = \lambda \sum_{k=1}^{K} |\theta_k|$$

with $\lambda > 0$. If we want to decrease the magnitude of the parameters, we would use **ridge regression:**

$$REGULARIZER(\boldsymbol{\theta}) = \lambda \sum_{k=1}^{K} {\theta_k}^2$$

again with $\lambda > 0$. The final objective of the ML algorithm for regression is then

$$\min_{\boldsymbol{\theta}} \sum_{t=1}^{T} (f(\mathbf{x}_t, \boldsymbol{\theta}) - y_t)^2 + REGULARIZER(\boldsymbol{\theta})$$

The **elastic net** combines the two. Nagel [13] discusses applications of regularization to asset pricing.

2.3 ▪ Hyperparameters and Validation Sets

Most ML algorithms have **hyperparameters**, which are parameters that are set by the user. For instance, we may want to regress a dependent variable on either affine, or linear, or quadratic functions in the independent variable. The degree of the polynomial is a hyperparameter.

For this reason, we introduce a **validation set**. While the training set was used to determine the optimal value of the parameters, the validation set is used to determine the **identity** of the parameters (i.e., what parameters should be used), or some other hyperparameters that depend on the algorithm (such as learning rate, tuning rate). When the dataset is not large enough, many techniques exist to mix and match the data into training set and validation set. This part of ML is called **cross-validation**.

2.4 ▪ Supervised Learning Algorithms

2.4.1 ▪ Probabilistic Supervised Learning

Most supervised learning algorithms can be viewed as estimating a conditional probability distribution $\pi(y|\boldsymbol{x};\boldsymbol{\theta})$, where $\boldsymbol{\theta}$ is a parameter. For instance, linear regression corresponds to the family

$$\pi(y|\boldsymbol{x};\boldsymbol{\theta}) = N(y, \boldsymbol{\theta}'\mathbf{x}, \mathbf{I})$$

where N is the normal probability distribution. When y takes value 0 or 1, we can use **logistic regression**. Denoting by σ the logistic function defined in Chapter 11, we set our model to be

$$\pi(y = 1|\boldsymbol{x};\boldsymbol{\theta}) = \sigma(\boldsymbol{\theta}'\mathbf{x})$$

2.4.2 ▪ Support Vector Machines

Support vector machines are used for classification. For instance, they predict that if $\boldsymbol{\theta}'\ \boldsymbol{x} > \theta_0$, then $\mathbf{x}$ belongs to a class. Otherwise, it belongs to another class.

2.4.3 ▪ Decision Trees

Decision trees are algorithms that break the input space into regions. Regions are determined one feature at a time. That is, we first split the sample into two datasets, one containing all $\{\mathbf{x}_t\}$ such that (for instance, selecting the first feature) $x_{t,1} < \theta_1$ and the other containing all the others. Then we repeat the process for each subset using a different feature (not necessarily the same for both subsets), and so on. The thresholds θ_k are chosen so as to maximize the discrepancy of the targets between the two separated sets. The same feature can come back several times, as illustrated in Figure 2.4.

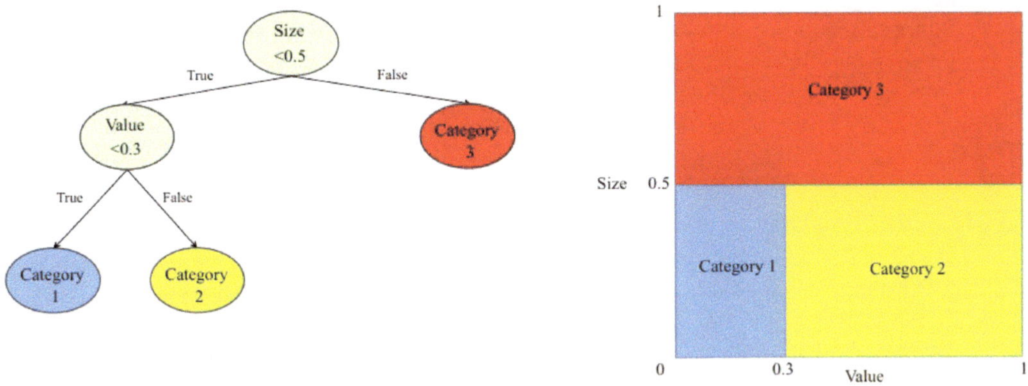

Figure 2.4: Regression tree example, from Figure 1 in [14]. The figure displays a regression tree (left) and its spatial representation based on "size" and "b/m" (right). Stocks are initially categorized by size, with those above a 0.5 breakpoint going to Category 3. Smaller stocks are further classified by b/m: below 0.3 to Category 1 and above to Category 2. The forecast for each category is determined by averaging the outcome variable within that group.

A **random forest** constructs a multitude of decision trees. **Boosting** is a meta-algorithm used in ML to give more weight to more relevant observations. The term is used mostly in the context of decision trees, but its principle underlies other algorithms, like Cochrane–Orcutt in generalized least squares and generative adversarial networks, which we will present in this book.

2.4.4 ▪ Neural Networks

Neural networks are currently one of the most powerful techniques in supervised learning. They are discussed extensively in Chapters 11, 12, and 13. When a neural network has several layers, it is called a **deep learning** algorithm.

2.5 ▪ Unsupervised Learning Algorithms

The classical problem of multivariate statistics, **statistical inference** can be considered part of unsupervised learning. This book will briefly review principles of **density estimation** before moving to nonparametric methods:

- principal component analysis (PCA)
- clustering.

These methods are so important that we dedicate a special chapter to them. As we shall see, the line separating supervised and unsupervised learning is often blurred.

Exercises

2.1. (Easy) Prove equation (2.1).

2.2. One of the most popular nonlinear regression algorithms currently is the deep feedforward neural network. This problem is meant to show that you can code your deep neural network yourself in less than one page. All explanations will be given in Chapter 11.[14] Suppose that a firm can have only two possible credit ratings: "good" and "bad." We are given a training set of 10 firms. The attributes of firm k are

$x_1(i)$	earnings/price of firm i
$x_2(i)$	volatility of firm i stock return

Construct a classifier that, for each new firm j, decides whether the firm is "good" or "bad," based on its earnings/price ratio and its volatility.
The training set is below:

i	1	2	3	4	5	6	7	8	9	10
$x_1(i)$	0.1	0.3	0.1	0.6	0.4	0.6	0.5	0.9	0.4	0.7
$x_2(i)$	0.1	0.4	0.5	0.9	0.2	0.3	0.6	0.2	0.4	0.6

You should implement the classifier as a deep neural network. The MATLAB code for it [37] is given below. You should

- either just run this program in MATLAB, or
- convert it into Python.

```
function netbp
%uses backpropagation to train network
%%%% DATA %%%%%%%%%%
    x1=[0.1,0.3,0.1,0.6,0.4,0.6,0.5,0.9,0.4,0.7];
    x2=[0.1, 0.4, 0.5, 0.9, 0.2, 0.3, 0.6, 0.2, 0.4, 0.6];
    y=[ones(1,5) zeros(1,5); zeros(1,5) ones(1,5)];
    % Initialize weights and biases
    rng(5000);
    W2=0.5*randn(2,2); W3=0.5*randn(3,2); W4=0.5*randn(2,3);
    b2=0.5*randn(2,1); b3=0.5*randn(3,1); b4=0.5*randn(2,1);
    %Forward and back propagate
    eta=0.05; %learning rate
    Niter=1e6; %number of Stochastic Gradient (SG) iterations
    savecost=zeros(Niter,1);
    for counter=1:Niter
        k=randi(10);
        x=[x1(k);x2(k)];
        % Forward pass
        a2=activate(x,W2,b2);
        a3=activate(a2,W3,b3);
        a4=activate(a3,W4,b4);
        %Backward pass
```

[14]This code is borrowed from [37].

```
            delta4=a4.*(1-a4).*(a4-y(:,k));
            delta3=a3.*(1-a3).*(transpose(W4)*delta4);
            delta2=a2.*(1-a2).*(transpose(W3)*delta3);
            %Gradient step
            W2=W2-eta*delta2*x';
            W3=W3-eta*delta3*a2';
            W4=W4-eta*delta4*a3';
            b2=b2-eta*delta2;
            b3=b3-eta*delta3;
            b4=b4-eta*delta4;
            %Monitor progress
            newcost=cost(W2,W3,W4,b2,b3,b4) %display cost to screen
            savecost(counter)=newcost;
        end
        %Show decay of cost function
        save costvec
        semilogy([1:1e4:Niter],savecost(1:1e4:Niter))
         function costval=cost(W2,W3,W4,b2,b3,b4)
            costvec=zeros(10,1);
            for i=1:10
                x=[x1(i); x2(i)];
                a2=activate(x,W2,b2);
                a3=activate(a2,W3,b3);
                a4=activate(a3,W4,b4);
                costvec(i)=norm(y(:,i)-a4,2);
            end
            costval=norm(costvec,2)^2;
         end % of nested function
    end

    function y=activate(x,W,b)
    %ACTIVATE Evaluates sigmoid function
    %
    % x is the input vector, y is is the output vector
    % W contains the weights, b contains the shifts
    %
    % The ith component of y is activate ((Wx+b)_i)
    % where activate =1/(1+exp(-z))
    y=1./(1+exp(-(W*x+b)));
```

Chapter 3

Classical Programming

The classical programming problem is to maximize a twice continuously differentiable function $f(\mathbf{x})$ under the equality constraints $\mathbf{g}(\mathbf{x}) = \mathbf{b}$:

$$\max_{\mathbf{x}} f(\mathbf{x})$$
$$\mathbf{g}(\mathbf{x}) = \mathbf{b}$$

where the variable vector $\mathbf{x} \in \mathbb{R}^N$ and $\mathbf{b} \in \mathbb{R}^M$. The function f is referred to as the *objective function*. An optimal solution to the problem, when it exists, is called a *global maximizer*. When it is unique, it is denoted by $\mathbf{x}^*$. The value $f(\mathbf{x}^*)$ is called the *maximum* of the problem. If one studies instead the problem of minimizing the objective, the solution will be called the *minimum* of the problem. The term *optimum* refers to the maximum (for a maximization problem) or the minimum (for a minimization problem).

The set of all $\mathbf{x}$ that satisfies the constraints is called the *feasible set* (or domain), or *opportunity set*. For the feasible set to be nonempty in general, the number of constraints should be less than the number of variables, i.e., $M \leq N$.

Example 3.1.

$$\max_{\mathbf{x}} -\frac{1}{2}\left[(x_1 - 4)^2 + (x_2 - 2.5)^2\right]$$
$$2x_2 - 3x_1 = 0$$

This example is illustrated in Figure 3.1, together with some *isocontours* or *level sets* of the function f, i.e., points $\mathbf{x}$, where $f(\mathbf{x})$ is constant.

We now determine conditions for a point $\mathbf{x}^*$ to solve the classical programming problem. We first consider the unconstrained case, familiar from calculus, before moving to the constrained case.

3.1 ▪ The Unconstrained Case

3.1.1 ▪ One Variable

Suppose that x^* is a local *maximizer*, i.e., a point such that, for some $\Delta x > 0$,

$$f(x^* + \eta\Delta x) \leq f(x^*) \quad \forall \eta \in [0, 1] \tag{3.1}$$

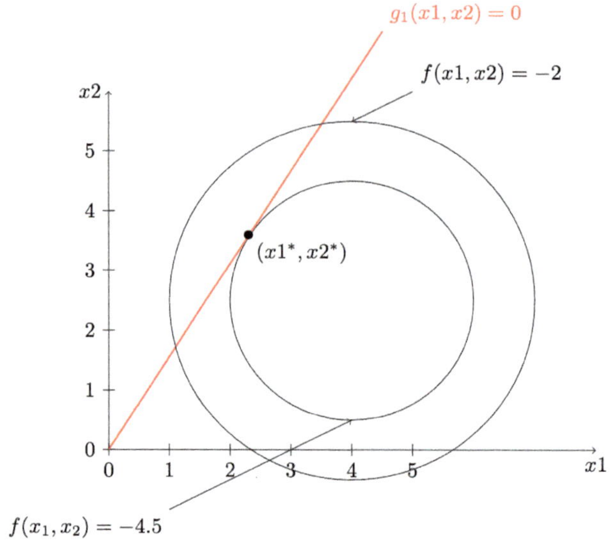

Figure 3.1: Example of classical programming.

and

$$f(x^* - \eta\Delta x) \leq f(x^*) \quad \forall \eta \in [0, 1] \tag{3.2}$$

Since f is assumed twice-differentiable, we can use the mean-value theorem. For each Δx, there exists a θ between 0 and 1 such that

$$f(x^* + \Delta x) = f(x^*) + \frac{df}{dx}(x^*)\Delta x + \frac{1}{2!}\frac{d^2 f}{dx^2}(x^* + \theta\Delta x)\Delta x^2 \tag{3.3}$$

Thus, for arbitrary small Δx we must have the *fundamental inequality*:

$$\frac{df}{dx}(x^*)\Delta x + \frac{1}{2}\frac{d^2 f}{dx^2}(x^* + \theta\Delta x)\Delta x^2 \leq 0 \tag{3.4}$$

We now divide (3.4) by Δx. If Δx is positive, we see that as Δx approaches zero,

$$\frac{df}{dx}(x^*) \leq 0$$

If Δx is negative, we see that as Δx approaches zero,

$$\frac{df}{dx}(x^*) \geq 0$$

Thus, for x^* to be a maximizer, the following *first-order condition* must hold:

$$\frac{df}{dx}(x^*) = 0 \tag{3.5}$$

Inserting (3.5) into (3.4), we obtain the following condition, valid for some $\theta \in [0, 1]$:

$$\frac{d^2 f}{dx^2}(x^* + \theta\Delta x) \leq 0$$

Now, we make Δx tend to zero. By continuity of the second-order derivative, we obtain the following facts.

Fact 3.2. *The first-order and second-order necessary conditions for x^* to be a local maximum are*

$$\frac{df}{dx}(x^*) = 0$$
$$\frac{d^2 f}{dx^2}(x^*) \le 0$$

When inequalities (3.1), (3.2) are replaced by strict inequalities, we obtain the following definition.

Definition 3.3. *A point x^* is a strict local maximizer if, for some $\Delta x > 0$, we have*

$$f(x^* + \eta\Delta x) < f(x^*) \quad \forall \eta \in (0, 1] \tag{3.6}$$

and

$$f(x^* - \eta\Delta x) < f(x^*) \quad \forall \eta \in (0, 1] \tag{3.7}$$

Fact 3.4. *Sufficient conditions for a strict local maximum are*

$$\frac{df}{dx}(x^*) = 0$$
$$\frac{d^2 f}{dx^2}(x^*) < 0$$

Different types of maxima/minima are illustrated in Figure 3.2.

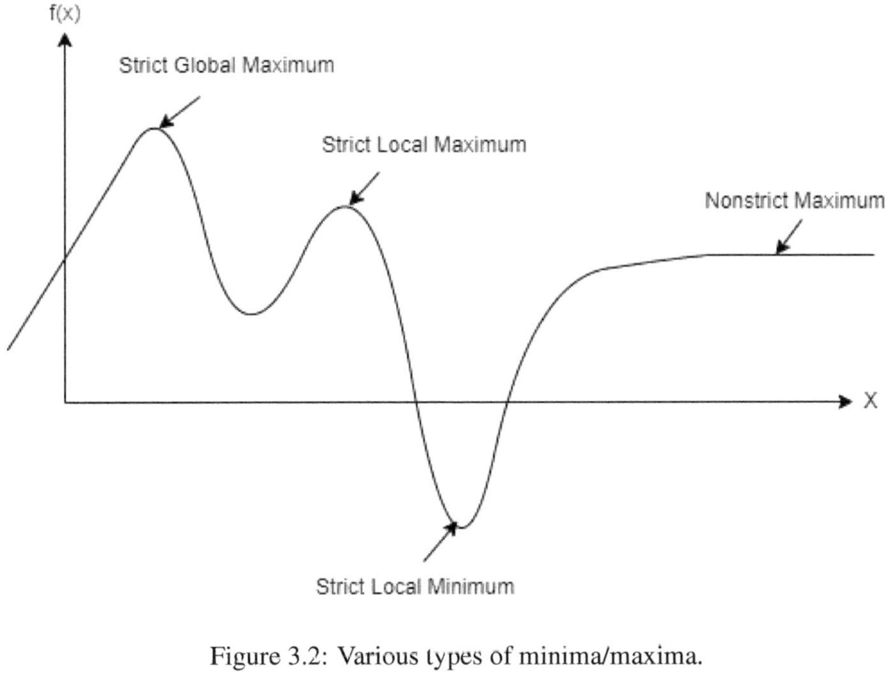

Figure 3.2: Various types of minima/maxima.

There are two strict local maximizers with different objective function values. How would a computer determine which one is the global maximizer? This problem of *global optimization* is a very difficult one and, unless one takes some assumptions on f, can be solved only by *brute search*, or exhaustive search. An analogy is a mountaineer trying to find the top of a mountain where there is fog, and she has no altimeter or GPS: at any point, she can only determine (by walking around) that she has reached a local maximum.

When a function is *concave*[15] it has only one strict maximum, so the problem is simpler. *Convex optimization* is the discipline that studies convex problems. Concavity and convexity are defined in the problem section.

3.1.2 ▪ The Multivariate Case

Suppose that $\mathbf{x}^*$ is a local maximizer. As before, for each $\mathbf{\Delta x}$, there exists a θ between -1 and 1 such that

$$f(\mathbf{x}^* + \mathbf{\Delta x}) = f(\mathbf{x}^*) + \frac{df}{d\mathbf{x}}(\mathbf{x}^*)\mathbf{\Delta x} + \frac{1}{2}\mathbf{\Delta x}^t \frac{d^2 f}{d\mathbf{x}^2}(\mathbf{x}^* + \theta\mathbf{\Delta x})\mathbf{\Delta x}$$

Thus, for arbitrary small $\mathbf{\Delta x}$ we must have

$$\frac{df}{d\mathbf{x}}(\mathbf{x}^*)\mathbf{\Delta x} + \frac{1}{2}\mathbf{\Delta x}^t \frac{d^2 f}{d\mathbf{x}^2}(\mathbf{x}^* + \theta\mathbf{\Delta x})\mathbf{\Delta x} \leq 0 \tag{3.8}$$

Like before, the fundamental inequality (3.8) is satisfied if $\mathbf{x}^*$ satisfies the *first-order* and *second-order* necessary conditions for a local maximum:

$$\frac{df}{d\mathbf{x}}(\mathbf{x}^*) = 0$$
$$\frac{d^2 f}{d\mathbf{x}^2}(\mathbf{x}^*) \leq 0$$

Definition 3.5. *A point that satisfies the first-order condition is called a stationary point, or a critical point.*

When only $\frac{d^2 f}{d\mathbf{x}^2}(\mathbf{x}^*) = 0$, $\mathbf{x}^*$ could be a nonstrict maximizer.[16] For $\mathbf{x}^*$ to be a strict local maximizer, one needs more.

Fact 3.6. *Sufficient conditions for a strict local maximum are*

$$\frac{df}{d\mathbf{x}}(\mathbf{x}^*) = 0$$
$$\frac{d^2 f}{d\mathbf{x}^2}(\mathbf{x}^*) < 0 \tag{3.9}$$

Condition (3.9) means that the Hessian (i.e., the matrix of second derivatives) must be strictly negative semidefinite, i.e., that all the eigenvalues must be strictly negative at the strict local maximizer $\mathbf{x}^*$. When $\frac{d^2 f}{d\mathbf{x}^2}(\mathbf{x}) < 0$ for all $\mathbf{x}$ in the domain, then $\mathbf{x}^*$ is a global maximizer.

[15] Or, more generally, when it is quasiconcave.

[16] One can construct pathological cases where the point $\mathbf{x}^*$ is not even a local maximizer.

3.2 ▪ The Constrained Case

The method of *Lagrange multipliers* shows that, at a local maximum, the gradient of the constraint and of the objective functions should be colinear (see Figure 3.3). This fact is very easy to visualize when there is only one constraint. For this reason, we present this case first.

3.2.1 ▪ One Constraint

We first take the simplest possible case, with only two variables. The problem to solve is

$$\max f(x_1, x_2)$$
$$g(x_1, x_2) = b$$

We write the differential of the constraint:

$$dg = \frac{\partial g}{\partial x_1} dx_1 + \frac{\partial g}{\partial x_2} dx_2 = 0$$

Assume that $\frac{\partial g}{\partial x_2}(x^*) \neq 0$. Thus, along the constraint, and in a small neighborhood of x^*,

$$\frac{dx_2}{dx_1} = \frac{-\frac{\partial g}{\partial x_1}}{\frac{\partial g}{\partial x_2}}$$

Define the curve $g(x_1, x_2) = b$ implicitly by $x_2 = h(x_1)$. We have

$$\frac{dh}{dx_1} = \frac{-\frac{\partial g}{\partial x_1}}{\frac{\partial g}{\partial x_2}}$$

We define

$$H(x_1) = f(x_1, h(x_1))$$

and we are back to the unconstrained case. We use Fact 3.2. The first-order conditions for $\mathbf{x}^*$ to be a maximizer are

$$\frac{dH}{dx_1}(\mathbf{x}^*) = 0 \tag{3.10}$$

We define the *Lagrange multiplier* y^* by

$$y^* = \frac{\frac{\partial f}{\partial x_2}(\mathbf{x}^*)}{\frac{\partial g}{\partial x_2}(\mathbf{x}^*)}$$

We develop the left-hand side of (3.10) and omit the argument $\mathbf{x}^*$ for clarity:

$$\begin{aligned}\frac{dH}{dx_1} &= \frac{\partial f}{\partial x_1} + \frac{\partial f}{\partial x_2}\frac{dh}{dx_1} \\ &= \frac{\partial f}{\partial x_1} - \frac{\frac{\partial f}{\partial x_2}}{\frac{\partial g}{\partial x_2}}\frac{\partial g}{\partial x_1} \\ &= \frac{\partial f}{\partial x_1} - y^*\frac{\partial g}{\partial x_1}\end{aligned}$$

Obviously, the following tautology holds:

$$\begin{aligned}0 &= \frac{\partial f}{\partial x_2} - \frac{\frac{\partial f}{\partial x_2}}{\frac{\partial g}{\partial x_2}}\frac{\partial g}{\partial x_2} \\ &= \frac{\partial f}{\partial x_2} - y^*\frac{\partial g}{\partial x_2}\end{aligned}$$

We see then that, for $j = 1, 2$, we must have

$$\frac{\partial f}{\partial x_j}(\mathbf{x}^*) - y^* \frac{\partial g}{\partial x_j}(\mathbf{x}^*) = 0$$

Thus the gradient of the constraint at the optimizer, namely, $\frac{dg}{d\mathbf{x}}(\mathbf{x}^*)$, must be parallel to the gradient of the objective function $\frac{df}{d\mathbf{x}}(\mathbf{x}^*)$, as Figure 3.3 illustrates.

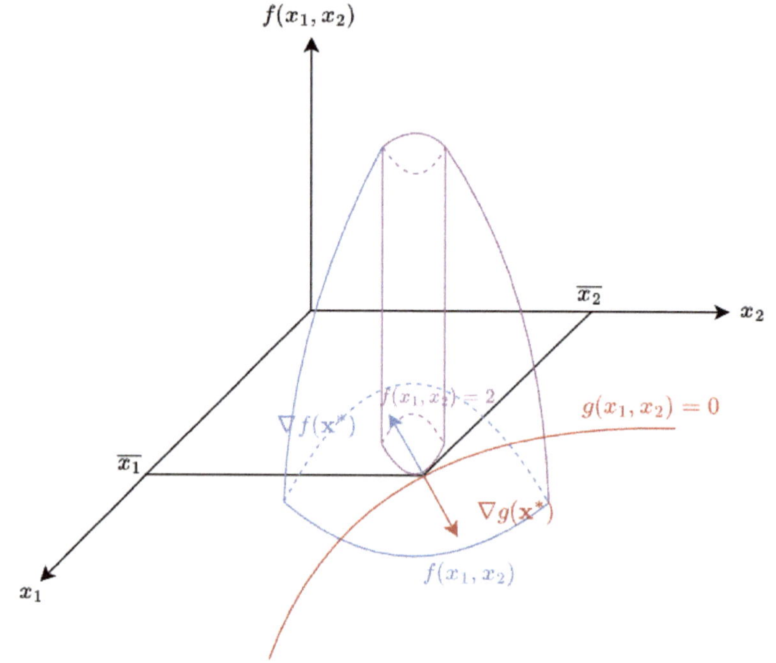

Figure 3.3: This figure illustrates the colinearity between the gradient of the objective and the gradient of the constraint at the solution.

3.2.2 ▪ Multiple Constraints

Let $M \leq N$. Assume that $\frac{d\mathbf{g}}{d\mathbf{x}}(\mathbf{x}^*)$ has full row rank. In a neighborhood of the solution $\mathbf{x}^*$, we parameterize the surface $\mathbf{b} = \mathbf{g}(\mathbf{x})$ by

$$\mathbf{x}^2 = \mathbf{h}(\mathbf{x}^1)$$

where $\mathbf{x}^1 \in \mathbb{R}^M$ and

$$\mathbf{x} = \begin{bmatrix} \mathbf{x}^1 \\ \mathbf{x}^2 \end{bmatrix}$$

Thus we can transform the classical programming problem into

$$\max H(\mathbf{x}^1) = \max f(\mathbf{x}^1, \mathbf{h}(\mathbf{x}^1))$$

We write the differential of the constraint in a neighborhood of the solution:

$$d\mathbf{g} = \frac{\partial \mathbf{g}}{\partial \mathbf{x}_1} d\mathbf{x}_1 + \frac{\partial \mathbf{g}}{\partial \mathbf{x}_2} \frac{d\mathbf{h}}{d\mathbf{x}_1} d\mathbf{x}^1 = 0$$

By the implicit function theorem,

$$\frac{d\mathbf{h}}{d\mathbf{x}_1} = -\left(\frac{\partial \mathbf{g}}{\partial \mathbf{x}_2}\right)^{-1} \frac{\partial \mathbf{g}}{\partial \mathbf{x}_1}$$

We define the *Lagrange multiplier* $\mathbf{y}^*$ by

$$\mathbf{y}^* = \frac{\partial f}{\partial \mathbf{x}_2}'(\mathbf{x}^*)\left(\frac{\partial \mathbf{g}}{\partial \mathbf{x}_2}(\mathbf{x}^*)\right)^{-1}$$

Omitting the argument $\mathbf{x}^*$, the gradient of H is

$$\begin{aligned} \frac{dH}{d\mathbf{x}_1} &= \frac{\partial f}{\partial \mathbf{x}_1} + \frac{\partial f}{\partial \mathbf{x}_2}' \frac{d\mathbf{h}}{d\mathbf{x}_1} \\ &= \frac{\partial f}{\partial \mathbf{x}_1} - \frac{\partial f}{\partial \mathbf{x}_2}' \left(\frac{\partial \mathbf{g}}{\partial \mathbf{x}_2}\right)^{-1} \frac{\partial \mathbf{g}}{\partial \mathbf{x}_1} \\ &= \frac{\partial f}{\partial \mathbf{x}_1} - \mathbf{y}\frac{\partial \mathbf{g}}{\partial \mathbf{x}_1} \end{aligned}$$

The other condition is a tautology:

$$\frac{\partial f}{\partial \mathbf{x}_2} - \frac{\partial f}{\partial \mathbf{x}_2}' \left(\frac{\partial \mathbf{g}}{\partial \mathbf{x}_2}\right)^{-1} \frac{\partial \mathbf{g}}{\partial \mathbf{x}_2} = 0$$

We can group these two conditions elegantly by defining the Lagrangian.

Definition 3.7. *The Lagrangian L is defined by*

$$L(\mathbf{x}, \mathbf{y}) = f(\mathbf{x}) + \mathbf{y}(\mathbf{b} - \mathbf{g}(\mathbf{x}))$$

Fact 3.8 (first-order optimality conditions). *The vector $\mathbf{x}^*$ is a local maximizer only if there is a row vector $\mathbf{y}^*$ such that $(\mathbf{x}^*, \mathbf{y}^*)$ is a stationary point of L.*

In other words, $(\mathbf{x}^*, \mathbf{y}^*)$ must satisfy

$$\frac{\partial L}{\partial \mathbf{x}}(\mathbf{x}^*) = \frac{df}{d\mathbf{x}}(\mathbf{x}^*) - \mathbf{y}^* \frac{d\mathbf{g}}{d\mathbf{x}}(\mathbf{x}^*) = 0 \tag{3.11}$$

$$\frac{\partial L}{\partial \mathbf{y}}(\mathbf{x}^*) = \mathbf{b} - \mathbf{g}(\mathbf{x}^*) = 0 \tag{3.12}$$

Second-order optimality conditions can also be worked out. We refer the interested reader to the book by Bazaara and Shetty [16], as we will not need them in this book.

3.2.3 ▪ Interpretation of the Lagrange Multipliers

You will show in the exercises that the Lagrange multipliers $\mathbf{y}^*$ are the sensitivity of $f(\mathbf{x}^*)$ to variations in $\mathbf{b}$:

$$\mathbf{y}^* = \frac{df}{d\mathbf{b}}(\mathbf{x}^*(\mathbf{b})) \tag{3.13}$$

An economist would say that y_m is the **marginal cost** of each resource m with resource capacity b_m. Marginal cost shows how the overall cost f increases with increased use of the resource b_m. We will return to this fundamental observation in our Chapter 5 on linear programming and build on it in Chapter 21, where we will show that the Lagrange multiplier for a portfolio problem is the state price. Increasing the state price will result in a commensurate increase in the price of an asset.

Exercises

3.1. (Easy) Solve graphically and analytically the following optimization problems:
(1) $\min x^2 - 2x + 1$
(2) $\min x^4 - 2x^2 + 1$
(3) $\min \exp(x + 1)$
(4) $\max \ln(\max(1, 5 + (2 - x)^2)$

3.2. (Medium) A function $f(\mathbf{x})$ is **convex** if, for each $\lambda \in (0, 1)$ and $\mathbf{x}_1, \mathbf{x}_2$,

$$f(\lambda \mathbf{x}_1 + (1 - \lambda)\mathbf{x}_2) \leq \lambda f(\mathbf{x}_1) + (1 - \lambda) f(\mathbf{x}_2)$$

Show that twice differentiable function f is convex if and only if, for any $\mathbf{x}, \bar{\mathbf{x}}$,

$$f(\mathbf{x}) \geq f(\bar{\mathbf{x}}) + \nabla f(\bar{\mathbf{x}})'(\mathbf{x} - \bar{\mathbf{x}})$$

Conclude that $\mathbf{x}^*$ is a global minimizer of f if the first-order condition is satisfied. This fact explains why we do not emphasize second-order conditions in this book.

3.3. (Medium) Consider the problem to maximize $\mathbf{x}'\mathbf{A}\mathbf{x}$, where $\mathbf{A}$ is a symmetric 2-by-2 non-diagonal matrix with $||\mathbf{x}|| = 1$. Construct examples where $\mathbf{A}$ is (a) positive definite, (b) negative definite, (c) neither, and indicate the optimal solution in each case.
Hint: Use the spectral decomposition of $\mathbf{A} = \mathbf{U}\boldsymbol{\Lambda}\mathbf{U}^{-1}$, where $\boldsymbol{\Lambda}$ is a diagonal matrix of eigenvalues and the matrix $\mathbf{U}$ is orthonormal.

3.4. (Medium) The multivariate Gaussian probability density function in I dimensions is

$$f(\mathbf{x}) = \frac{1}{\sqrt{(2\pi)^I |\boldsymbol{\Sigma}|}} \exp\left(-\frac{1}{2}(\mathbf{x} - \mathbf{m})' \boldsymbol{\Sigma}^{-1} (\mathbf{x} - \mathbf{m})\right)$$

Given $\{\mathbf{x}_t\}$ samples for $t = 1, \ldots, T$, the **maximum likelihood method** seeks to find the parameters $\mathbf{m}$ (the mean) and $\boldsymbol{\Sigma}$ (the variance-covariance matrix) that maximize the likelihood function

$$Lik(\mathbf{m}, \boldsymbol{\Sigma}) = \prod_{t=1}^{T} f(\mathbf{x}_t; \mathbf{m}, \boldsymbol{\Sigma})$$

Show, by maximizing the logarithm of the likelihood, that the maximum likelihood estimator of the mean is the sample mean:

$$\mathbf{m}^* = \frac{1}{T}\sum_{t=1}^{T} \mathbf{x}_t$$

What is the maximum likelihood estimator of the variance-covariance matrix?

3.5. (Medium) **Linear Regression**
You are given the following data (x_t, y_t) for $t = 1, \ldots, T$. The problem is to fit this data by minimizing the *residual sum of squares* (RSS):

$$\min_{a,b} RSS(a, b) = \min \sum_{t=1}^{T} (y_t - (a + bx_t))^2$$

Determine the necessary conditions for minimizing $RSS(a, b)$ by choice of a and b. These equations are called the *normal equations*. Show that the sufficient conditions are met.

3.6. (Medium) In the classical programming problem, write $\mathbf{x}^* = \mathbf{x}^*(\mathbf{b})$ and $\mathbf{y}^* = \mathbf{y}^*(\mathbf{b})$. Show that (3.13) holds.

3.7. (Hard) Let $\mathbf{c}$ be a row vector. Consider the linear quadratic programming problem

$$\begin{aligned} \max_{\mathbf{x}} \mathbf{c}\mathbf{x} + \frac{1}{2}\mathbf{x}'\mathbf{D}\mathbf{x} \\ \mathbf{A}\mathbf{x} = \mathbf{b} \end{aligned}$$

where $\mathbf{D}$ is negative definite and $\mathbf{A}\mathbf{D}^{-1}\mathbf{A}'$ is invertible. The latter condition is generically satisfied when there are less constraints than variables.

(a) Set up the Lagrangian, and write the first-order conditions.

(b) Solve for the Lagrange multiplier $\mathbf{y}^*$ as a function of $\mathbf{A}, \mathbf{D}, \mathbf{b}, \mathbf{c}$.

(c) Solve for the optimal solution $\mathbf{x}^*$, first as a function of the Lagrange multiplier and then, using part (b), as a function of $\mathbf{A}, \mathbf{D}, \mathbf{b}, \mathbf{c}$.

(d) Obtain the sensitivities $\frac{\partial \mathbf{x}^*}{\partial \mathbf{b}}$ and $\frac{\partial \mathbf{x}^*}{\partial \mathbf{c}}$, and verify that $\mathbf{x}^*$ varies linearly as the parameters change.

3.8. (Easy) Solve graphically using contours and opportunity sets the following problems:

(a) $\max x_1 + x_2$ subject to $x_1^2 + x_2^2 = 1$.

(b) $\max \exp(-(x_1^2 + 2x_2^2))$ subject to $2x_1 + 3x_2 = 4$.

(c) $\max \sin x_1 \cos x_2$ subject to $x_1 - x_2 = 0$.

3.9. (Medium) **Principal Component Analysis (PCA)**
Consider the following problem:

$$\begin{aligned} \max \mathbf{x}'\mathbf{A}\mathbf{x} \\ \mathbf{x}'\mathbf{x} = 1 \end{aligned}$$

where $\mathbf{A}$ is a given variance-covariance matrix. Show that if $\mathbf{x}^*$ is the solution, then $\mathbf{x}^{*\prime}\mathbf{A}\mathbf{x}^*$ is the largest eigenvalue of $\mathbf{A}$.

3.10. (Medium) **Right Moore–Penrose Inverse**
Suppose the $\mathbf{A}$ matrix has more columns than rows. Use Exercise 3.7 to show that the vector $\mathbf{x}^*$ of minimum Euclidean norm that solves $\mathbf{A}\mathbf{x} = \mathbf{b}$ is given by

$$\mathbf{x} = \mathbf{A}^+\mathbf{b}$$

where $\mathbf{A}^+ = \mathbf{A}'(\mathbf{A}\mathbf{A}')^{-1}$.

Chapter 4
Nonlinear Programming

The nonlinear programming problem (also referred to as *nonlinear program*) is

$$\max_{\mathbf{x}} f(\mathbf{x}) \tag{4.1}$$

$$\mathbf{g}(\mathbf{x}) \leq \mathbf{b} \tag{4.2}$$

$$\mathbf{x} \geq 0 \tag{4.3}$$

We will refer to inequalities (4.2) as *inequality constraints*, and constraints (4.3) as *positivity constraints*.

Example 4.1.

$$\begin{aligned} \max -\frac{1}{2}\left[(x_1-4)^2+(x_2-2.5)^2\right] \\ g_1(x_1,x_2) = 2x_2 - 3x_1 \leq 0 \\ g_2(x_1,x_2) = x_2 \leq 1.5 \\ x_1 \geq 0 \\ x_2 \geq 0 \end{aligned}$$

This example is illustrated in Figure 4.1.

The first goal of this chapter is to develop first-order optimality conditions, called *(Karush–) Kuhn–Tucker conditions*, like in the previous chapter. However, ML algorithms rarely use these results directly since they try to avoid modeling constraints explicitly: this fact motivates the role of regularizers and the role of the sigmoid function within a neuron. Both limit the possible values a variable can take.

The second goal of this chapter is to develop duality theory. To each original ("primal") problem corresponds a dual problem, whose optimum in the convex case corresponds to the optimum of the primal. More importantly, most algorithms that exploit duality can recover the optimizer of the primal problem from the optimizer of the dual problem. Some algorithms, like the primal-dual algorithm, alternate between solving a primal and a dual problem. While a full description of these algorithms is beyond the scope of the book, we will retain the following messages from duality theory:

- The idea of alternating between solving two different problems has inspired related algorithms in ML. We will see in Chapter 7 the Cochrane–Orcutt method, also used in general-

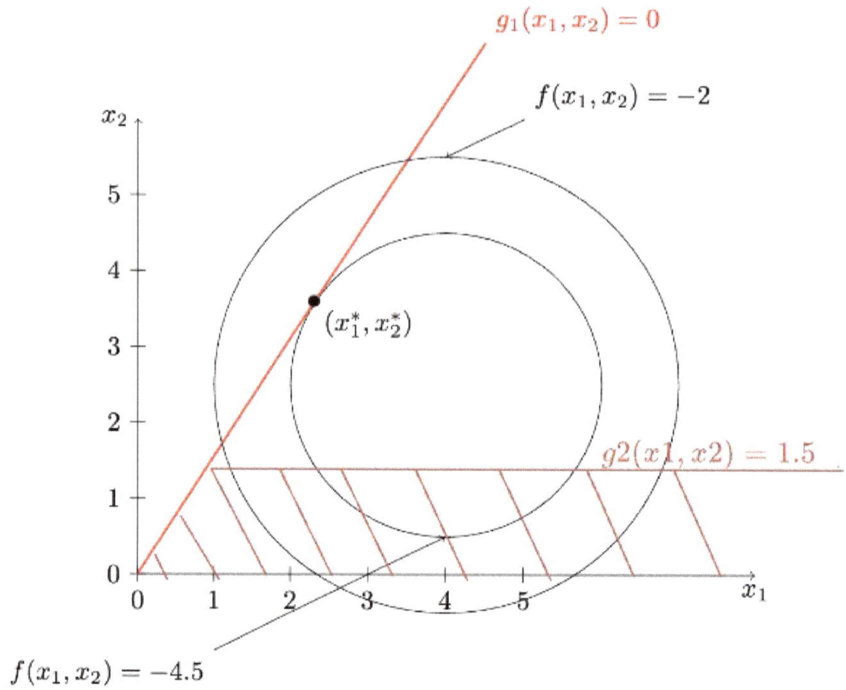

Figure 4.1: The feasible domain is in dark red. The optimum is as $(x_1^*, x_2^*) = (4, 1.5)$.

ized method of moments (GMM). Generative adversarial networks (GAN) in Chapter 13 also exploit this feature.

- In finance, there is a duality between the portfolio problem and the asset pricing problem, as we shall make clear in Chapter 21. This fact inspired many advanced algorithms in portfolio optimization, which again are beyond the scope of the book.

The reader more interested in applications than theory should skip this chapter and may want to memorize only the following (grossly simplified) conclusion.

(Simplified) conclusion: the Kuhn–Tucker conditions in nonlinear programming are essentially the same as the first-order conditions in classical programming, with the addition that Lagrange multipliers should be positive.

We start by omitting the inequality constraints from the problem and then move to the general problem. We skip technical considerations. For instance, we assume that the constraint qualification is always met.

4.1 ▪ The Case of No Inequality Constraints $\mathrm{g}(\mathrm{x}) \leq \mathrm{b}$

The problem becomes simply

$$\max_{\mathbf{x}} f(\mathbf{x})$$
$$\mathbf{x} \geq 0$$

Suppose that $\mathbf{x}^*$ is a local solution. This means that for all $\boldsymbol{\Delta}\mathbf{x}$ sufficiently small there exists a value of θ with $0 < \theta < 1$ such that

$$\frac{df}{d\mathbf{x}}(\mathbf{x}^*)^t \boldsymbol{\Delta}\mathbf{x} + \frac{1}{2}\boldsymbol{\Delta}\mathbf{x}^t \frac{d^2 f}{d\mathbf{x}^2}(\mathbf{x}^* + \theta\boldsymbol{\Delta}\mathbf{x})\boldsymbol{\Delta}\mathbf{x} \leq 0 \tag{4.4}$$

We will henceforth focus only on the first-order conditions; that is, we need to set

$$\frac{df}{d\mathbf{x}}(\mathbf{x}^*)^t \boldsymbol{\Delta}\mathbf{x} \leq 0$$

We investigate under what conditions (4.1) holds. Let us work variable by variable and consider the variable x_j^*. There are two cases:

1. The constraint is not binding, i.e., $x_j^* > 0$; since Δx_j can be positive or negative, (4.1) imposes $\frac{\partial f}{\partial x_j}(\mathbf{x}^*) = 0$.

2. The constraint is binding, or *tight*, i.e., $x_j^* = 0$, so that $\Delta x_j > 0$; optimality imposes $\frac{df}{dx_j}(\mathbf{x}^*) \leq 0$.

Suppose the first case holds, i.e., $x_j^* > 0$ for some j. The corresponding optimality condition ($\frac{\partial f}{\partial x_j}(\mathbf{x}^*) = 0$) clearly holds if $\frac{\partial f}{\partial x_j}(\mathbf{x}^*)x_j^* = 0$, $\frac{\partial f}{\partial x_j}(\mathbf{x}^*) \leq 0$. We now show that $\sum_{n=1}^N \frac{\partial f}{\partial x_n}(\mathbf{x}^*)x_n^* = 0$, together with $\frac{\partial f}{\partial x_i}(\mathbf{x}^*) \leq 0$ and $x_i^* \geq 0$ $(i \neq j)$, results in $\frac{\partial f}{\partial x_j}(\mathbf{x}^*)x_j^* = 0$. Indeed $\sum_{i \neq j} \frac{\partial f}{\partial x_i}(\mathbf{x}^*)x_i^* \leq 0$, so that

$$0 = \sum_{n=1}^{N} \frac{\partial f}{\partial x_n}(\mathbf{x}^*)x_n^* = \underbrace{\sum_{i \neq j} \frac{\partial f}{\partial x_i}(\mathbf{x}^*)x_i^*}_{\leq 0} + \frac{\partial f}{\partial x_j}(\mathbf{x}^*)x_j^*$$

Since $x_j^* > 0$ and $\frac{\partial f}{\partial x_j}(\mathbf{x}^*) \leq 0$ the only possibility is $\frac{\partial f}{\partial x_j}(\mathbf{x}^*) = 0$.
The second case is immediate. We thus proved the following fact.

Fact 4.2. *The vector* $\mathbf{x}^*$ *is a local maximizer only if*

$$\frac{df}{d\mathbf{x}}(\mathbf{x}^*) \leq 0 \tag{4.5}$$

$$\frac{df}{d\mathbf{x}}(\mathbf{x}^*)'\mathbf{x}^* = 0 \tag{4.6}$$

$$\mathbf{x}^* \geq 0 \tag{4.7}$$

Observe that these equations mean, component by component,

$$\frac{\partial f}{\partial x_n}(\mathbf{x}^*) \leq 0 \quad n = 1, \ldots, N$$

$$\sum_{n=1}^{N} \frac{\partial f}{\partial x_n}(\mathbf{x}^*)x_n^* = 0$$

$$x_n^* \geq 0 \quad n = 1, \ldots, N$$

4.2 ▪ The Karush–Kuhn–Tucker (KKT) Conditions

We consider now the general nonlinear program (4.1) to (4.3). The idea is to transform it into the problem of the last section, together with equality constraints, and reuse the Lagrangian theory used in Chapter 3. To that effect, we define *slack variables*:

$$\mathbf{s} = \mathbf{b} - \mathbf{g}(\mathbf{x})$$

The nonlinear program (4.1) to (4.3) becomes

$$\max_{\mathbf{x},\mathbf{s}} f(\mathbf{x}) \tag{4.8}$$

$$\mathbf{g}(\mathbf{x}) + \mathbf{s} = \mathbf{b} \tag{4.9}$$

$$\mathbf{x} \geq 0; \mathbf{s} \geq 0 \tag{4.10}$$

Define the Lagrangian:

$$L'(\mathbf{x}, \mathbf{y}, \mathbf{s}) = f(\mathbf{x}) + \mathbf{y}(\mathbf{b} - \mathbf{g}(\mathbf{x}) - \mathbf{s})$$

We now consider the following problem:

$$\max_{\mathbf{x},\mathbf{y},\mathbf{s}} L'(\mathbf{x}, \mathbf{y}, \mathbf{s}) \tag{4.11}$$

$$\mathbf{x} \geq 0; \mathbf{s} \geq 0 \tag{4.12}$$

Using Fact 4.2, we have the following necessary conditions for $(\mathbf{x}^*, \mathbf{y}^*, \mathbf{s}^*)$ to solve (4.11) and (4.12):

- optimality constraints for the variable $\mathbf{x}$; since it has positivity constraints, we use (4.5), (4.6), (4.7):

$$\frac{\partial L'}{\partial \mathbf{x}} \leq 0 \quad \frac{\partial L'}{\partial \mathbf{x}}\mathbf{x}^* = 0 \quad \mathbf{x}^* \geq 0$$

- optimality constraints for the variable $\mathbf{s}$; since it has positivity constraints, we use (4.5), (4.6), (4.7):

$$\frac{\partial L'}{\partial \mathbf{s}} \leq 0 \quad \frac{\partial L'}{\partial \mathbf{s}}\mathbf{s}^* = 0 \quad \mathbf{s}^* \geq 0.$$

- optimality constraints for the variable $\mathbf{y}$; since we do not know (yet) that it should be positive, we use $\frac{\partial L'}{\partial \mathbf{y}} = 0$.

By the Lagrangian theory seen in Chapter 3, necessary conditions for $(\mathbf{x}^*, \mathbf{y}^*, \mathbf{s}^*)$ to solve problem (4.11), (4.12) are necessary conditions for $\mathbf{x}^*$ to solve the general nonlinear program (4.1) to (4.3). Substituting the definition of L', and using (4.9) to get rid of the slack variable, we obtain the following KKT conditions.

Fact 4.3. *Necessary conditions for* $\mathbf{x}^*$ *to solve the nonlinear program for (4.1) to (4.3) are that, for some vector* $\mathbf{y}^*$, *the following KKT conditions are satisfied:*

$$\frac{df}{d\mathbf{x}} - \mathbf{y}^* \frac{d\mathbf{g}}{d\mathbf{x}} \leq 0 \quad \left(\frac{\partial L'}{\partial \mathbf{x}} \leq 0\right) \tag{4.13}$$

$$\left(\frac{df}{d\mathbf{x}} - \mathbf{y}^* \frac{d\mathbf{g}}{d\mathbf{x}}\right)' \mathbf{x}^* = 0 \quad \left(\frac{\partial L'}{\partial \mathbf{x}}\mathbf{x}^* \leq 0\right) \tag{4.14}$$

$$\mathbf{x}^* \geq 0 \tag{4.15}$$

$$\mathbf{y}^* \geq 0 \quad \left(\frac{\partial L'}{\partial \mathbf{s}} \leq 0\right) \tag{4.16}$$

$$\mathbf{y}^*(\mathbf{b} - \mathbf{g}(\mathbf{x}^*)) = 0 \quad \left(\frac{\partial L'}{\partial \mathbf{s}}\mathbf{s}^* \leq 0\right) \tag{4.17}$$

$$\mathbf{b} - \mathbf{g}(\mathbf{x}^*) \geq \mathbf{0} \quad \left(\mathbf{s}^* \geq 0 \text{ and } \frac{\partial L'}{\partial \mathbf{y}} = 0\right) \tag{4.18}$$

4.3 ▪ The Nonlinear Programming Problem without Positivity Constraints

The reader may be more familiar with the following problem:

$$\max f(\mathbf{x}) \tag{4.19}$$

$$\tilde{\mathbf{g}}(\mathbf{x}) \leq 0 \tag{4.20}$$

Clearly, (4.19) is a special case of (4.1). With elementary transformations, though, we can show that any program of the form (4.1) can be transformed into a program of the form (4.19). Indeed, set

$$\tilde{\mathbf{g}}(\mathbf{x}) = \begin{pmatrix} \mathbf{g}(\mathbf{x}) - \mathbf{b} \\ -\mathbf{x} \end{pmatrix}$$

Thus these two programs are equivalent. The KKT conditions take an easier form under (4.19). It is, however, easier to prove the KKT conditions for (4.1). For (4.19), we need to take an (interesting) detour through convex analysis. Interested readers should consult [16].

Fact 4.4. *The KKT conditions for (4.19), (4.20) are*

$$\frac{df}{d\mathbf{x}}(\mathbf{x}^*) - \boldsymbol{\lambda}^* \frac{d\tilde{\mathbf{g}}}{d\mathbf{x}}(\mathbf{x}^*) = 0 \tag{4.21}$$

$$\boldsymbol{\lambda}^* \tilde{\mathbf{g}}(\mathbf{x}^*) = 0 \tag{4.22}$$

$$\tilde{\mathbf{g}}(\mathbf{x}^*) \leq 0 \tag{4.23}$$

$$\boldsymbol{\lambda}^* \geq 0 \tag{4.24}$$

Equation (4.23) is called the *complementary slackness condition*: at least one of the multipliers or the constraint is equal to zero.

Thus the key differences with the optimality conditions in classical programming (refer to Fact 3.8) are

- the complementary slackness condition
- positivity of the multiplier.

The gradients of the constraint and of the objective must be colinear in both cases, as (3.11) shows for the classical programming case and (4.21) shows for the nonlinear programming case. A simple look at Figure 3.2 should be sufficient for the reader to remember that the multiplier λ^* is positive in the nonlinear programming case.

4.4 ▪ Saddle Points

Let $L(\mathbf{x},\mathbf{y}) = f(\mathbf{x}) + \mathbf{y}(\mathbf{b} - \mathbf{g}(\mathbf{x}))$. We can reorganize the Kuhn–Tucker conditions (4.13) to (4.18) into the necessary conditions for a maximum in Table 4.1.

(4.13)$\frac{\partial L}{\partial \mathbf{x}}(\mathbf{x}^*,\mathbf{y}^*) \leq 0$	(4.18)$\frac{\partial L}{\partial \mathbf{y}}(\mathbf{x}^*,\mathbf{y}^*) \geq 0$
(4.14)$\frac{\partial L}{\partial \mathbf{x}}(\mathbf{x}^*,\mathbf{y}^*)\mathbf{x}^* = 0$	(4.17)$\frac{\partial L}{\partial \mathbf{y}}(\mathbf{x}^*,\mathbf{y}^*)\mathbf{y}^* = 0$
(4.15)$\mathbf{x}^* \geq 0$	(4.16) $\mathbf{y}^* \geq 0$

(4.25)

Table 4.1: KKT conditions, classified so as to illustrate the saddle point property.

Observe the symmetry between the left and the right columns. Indeed, the only difference is that, in the first row, the derivative with respect to $\mathbf{x}$ is negative, while the derivative with respect to $\mathbf{y}$ is negative. This points to the fact that (provided the second-order conditions are met) $\mathbf{x}^*$ is a local maximum for the problem of maximizing $L(\mathbf{x},\mathbf{y})$ under positivity constraints, while $\mathbf{y}^*$ is a local minimum of a different problem under positivity constraints.

This formulation inspired the idea of a saddle point $(\mathbf{x}^*,\mathbf{y}^*)$, i.e., a point that is the maximum in the $\mathbf{x}$ direction, while being a minimum in the $\mathbf{y}$ direction.

Definition 4.5. *A pair $(\mathbf{x}^*,\mathbf{y}^*)$ is a saddle point if $(\mathbf{x}^*,\mathbf{y}^*)$ is nonnegative, and for all $\mathbf{x},\mathbf{y} \geq \mathbf{0}$,*

$$L(\mathbf{x},\mathbf{y}^*) \leq L(\mathbf{x}^*,\mathbf{y}^*) \leq L(\mathbf{x}^*,\mathbf{y})$$

Theorem 4.6 (saddle point theorem). *If $(\mathbf{x}^*,\mathbf{y}^*)$ is a saddle point, then $\mathbf{x}^*$ solves (4.1) to (4.3).*

See Figures 4.2 and 4.3 for an illustration of the saddle point in two cases.

Proof. The inequality $L(\mathbf{x}^*,\mathbf{y}^*) \leq L(\mathbf{x}^*,\mathbf{y})$ spells

$$f(\mathbf{x}^*) + \mathbf{y}^*(\mathbf{b} - \mathbf{g}(\mathbf{x}^*)) \leq f(\mathbf{x}^*) + \mathbf{y}(\mathbf{b} - \mathbf{g}(\mathbf{x}^*))$$

Subtracting the left term yields

$$(\mathbf{y} - \mathbf{y}^*)(\mathbf{b} - \mathbf{g}(\mathbf{x}^*)) \geq 0$$

Taking $\mathbf{y} \to \infty$ we see that

$$\mathbf{b} - \mathbf{g}(\mathbf{x}^*) \geq 0$$

Thus $\mathbf{x}^*$ is feasible. Taking $\mathbf{y} = \mathbf{0}$ we have $\mathbf{y}^*(\mathbf{b} - \mathbf{g}(\mathbf{x}^*)) \leq 0$. Thus, since both terms are positive,

$$\mathbf{y}^*(\mathbf{b} - \mathbf{g}(\mathbf{x}^*)) = 0$$

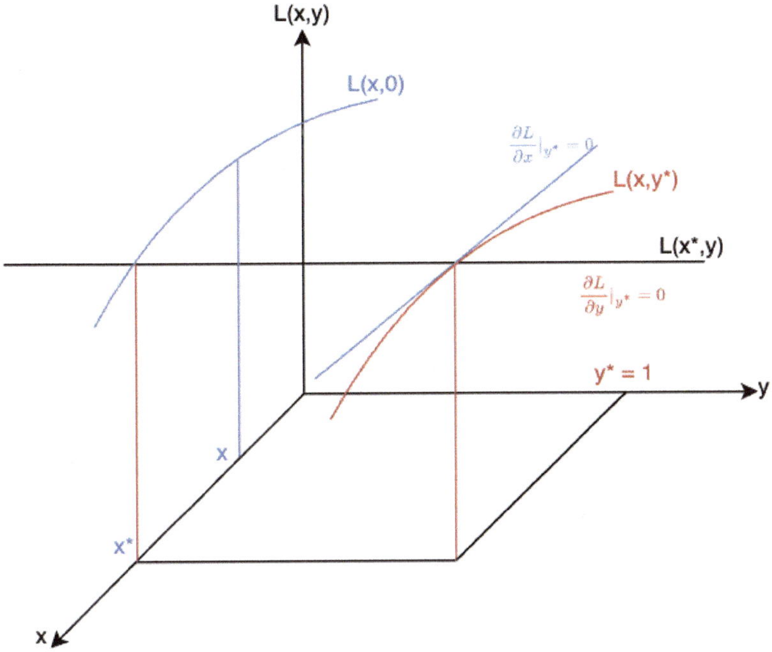

Figure 4.2: Saddle point when the constraint is tight at the optimum of the primal.

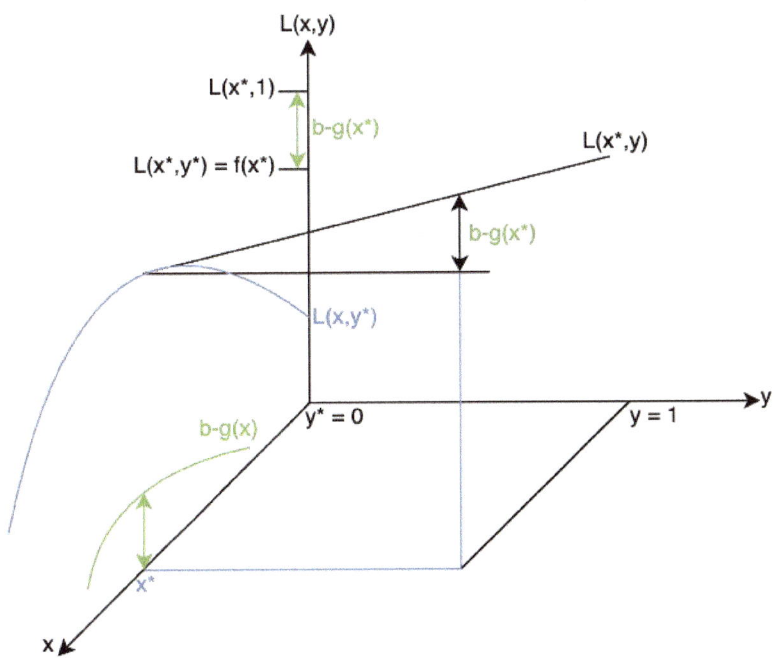

Figure 4.3: Saddle point when the constraint is not tight at the optimum of the primal.

As a consequence, the inequality $L(\mathbf{x}, \mathbf{y}^*) \leq L(\mathbf{x}^*, \mathbf{y}^*)$ becomes

$$f(\mathbf{x}) + \mathbf{y}^*(\mathbf{b} - \mathbf{g}(\mathbf{x})) \leq f(\mathbf{x}^*)$$

By positivity of both $\mathbf{y}^*$ and $\mathbf{b} - \mathbf{g}(\mathbf{x})$, we obtain $f(\mathbf{x}) \leq f(\mathbf{x}^*)$. ■

Note: Theorem 4.6 does not require convexity assumption for $\mathbf{x}^*$ to be a global maximizer of (4.1). However, the necessary conditions (4.25) are not sufficient for $\mathbf{x}^*$ to be a global maximizer (4.1). This "superiority" of Theorem 4.6 inspired researchers to solve the *dual problem* instead:

$$\max_{\mathbf{x}\geq\mathbf{0}} \min_{\mathbf{y}\geq\mathbf{0}} f(\mathbf{x}) + \mathbf{y}(\mathbf{b} - \mathbf{g}(\mathbf{x}))$$

whose solution, denoted by $(\mathbf{x}^*, \mathbf{y}^*)$, is clearly a saddle point and, by Theorem 4.6, is a global solution of (4.1). Unfortunately, the dual problem is, in general, a more complicated problem than the primal problem since it is an optimization problem over more variables ($\mathbf{x}$ and $\mathbf{y}$) than the primal (which has only $\mathbf{x}$). As we shall see, however, when the primal program is linear, the dual program will be formulated in terms of the dual variable $\mathbf{y}$ only.

One of the strongest results of mathematical finance consists in transforming the portfolio maximization problem into a probability measure minimization problem by using this duality result.

4.5 ▪ Other Formulations of the Nonlinear Programming Problem

You will see in the exercises that by appropriate transformations, the following problems can be in the form (4.1). The problem (4.19) is just one of them. Here are some others:

Minimization:

$$\begin{aligned} &\min_{\mathbf{x}} h(\mathbf{x}) \\ &\mathbf{g}(\mathbf{x}) \leq \mathbf{b} \\ &\mathbf{x} \geq 0 \end{aligned} \tag{4.26}$$

Null right-hand side:

$$\begin{aligned} &\max_{\mathbf{x}} f(\mathbf{x}) \\ &\mathbf{h}(\mathbf{x}) \leq \mathbf{0} \\ &\mathbf{x} \geq 0 \end{aligned} \tag{4.27}$$

No positivity constraints on x:

$$\begin{aligned} &\max_{\mathbf{x}} f(\mathbf{x}) \\ &\mathbf{g}(\mathbf{x}) \leq \mathbf{b} \end{aligned} \tag{4.28}$$

This problem is different from (4.19), (4.20) because we do not assume that $\mathbf{x} \geq \mathbf{0}$ is "hidden" in part of the constraint $\mathbf{g}(\mathbf{x}) \leq \mathbf{b}$. So negative solutions are allowed.

Equality constraints:

$$\begin{aligned} &\max_{\mathbf{x}} f(\mathbf{x}) \\ &\mathbf{h}(\mathbf{x}) = \mathbf{0} \\ &\mathbf{x} \geq 0 \end{aligned} \tag{4.29}$$

4.6 ▪ Notes

This material is classical and follows [17]. A more comprehensive treatment is given by [16]. Linear programming is not the only case where the dual problem is expressed only in terms of the dual variable $\mathbf{y}$. It is also the case when the primal problem is concave and there is only one constraint. In this case, one uses Fenchel–Legendre duality; see Exercise 4.10. This has been explored by many researchers in portfolio theory. See [18] for more details.

Exercises

4.1. (Easy) Solve graphically and analytically

$$\max x_1^2 + x_2^2$$
$$x_1 \geq 0$$
$$x_1 + x_2 \leq 3$$

Write the KKT optimality conditions.

4.2. (Easy) Derive Fact 4.4 from Fact 4.3.

4.3. (Easy) Define how to transform the minimization problem (4.26) into the nonlinear programming problem (4.1) to (4.3), so that they have the same solution(s).

4.4. (Easy) Define how to transform the null right-hand side program (4.27) into the nonlinear programming problem (4.1) to (4.3), so that they have the same solution(s).

4.5. (Easy) Define how to transform the no-positivity constraint program (4.28) into the nonlinear programming problem (4.1) to (4.3), so that they have the same solution(s).

4.6. (Easy) Define how to transform the equality constraint program (4.29) into the nonlinear programming problem (4.1) to (4.3), so that they have the same solution(s).

4.7. (Easy) Consider the problem of allocating a scarce resource b between N different tasks. The objective function is separable, and the problem is

$$\max_{\mathbf{x}} f_1(x_1) + \cdots + f_N(x_N)$$
$$x_1 + \cdots + x_N \leq b$$
$$\mathbf{x} \geq \mathbf{0}$$

Interpret the optimality conditions.

4.8. (Medium) Suppose that f is convex. Show that the problem of minimizing f under a linear constraint $\mathbf{ax} \geq b$ is the same as the problem of maximizing $\mathbf{ax}$ under a constraint of the form $f(\mathbf{x}) \leq c$. What should c be? For simplicity, suppose that the gradient of f at the optimum is nonzero.

We will see in the chapter on mean-variance that this has the following interpretation. Solving for the minimum variance portfolio under a constraint on the expected return is the same as solving for the maximum expected return portfolio under a constraint on the variance of the portfolio.

4.9. (Medium) Suppose f is convex. Investigate the relationship between the optimal solutions of the following two problems, where $\mathbf{y} \geq \mathbf{0}$ is a fixed vector:

$$\min_{\mathbf{x}} f(\mathbf{x}) \text{ s.t. } \mathbf{g}(\mathbf{x}) \leq 0$$

and

$$\min_{\mathbf{x}} f(\mathbf{x}) \text{ s.t. } \mathbf{y}'\mathbf{g}(\mathbf{x}) \leq 0$$

4.10. (Hard) **Legendre Duality**
Suppose f is a convex function. Consider the problem

$$\min_{\mathbf{x}} f(\mathbf{x})$$
$$\mathbf{x} \geq 0$$

The Legendre transform of f, also called the convex conjugate, is defined by

$$f^*(\mathbf{y}) = \max_{\mathbf{x}}(\mathbf{y}\mathbf{x} - f(\mathbf{x}))$$

One can call this problem the (Legendre) primal problem. Define the operation of Legendre transformation by the operator $\mathcal{L}$:

$$f^* = \mathcal{L}(f)$$

(a) Show that f is self-dual, i.e., $\mathcal{L}(\mathcal{L}(f)) = f$.

(b) Show that if $f(x) = cx^2$, then $f^*(y) = y^2/4c$.

(c) Find a microeconomic interpretation of the Legendre transform of the cost function $f(x)$, where x is quantity produced and y is the price of the product.

Chapter 5

Linear Programming

Linear programming is a special case of nonlinear programming, where the objective function and the constraints are linear. One of the main advantages of linear programming is its numerical robustness and speed. Linear programs with millions of variables and constraints are routinely solved in industry, and the convergence is guaranteed, unlike neural networks, for instance.

A linear program can be written as

$$\max_{\mathbf{x}} \mathbf{cx}$$
$$\mathbf{Ax} \leq \mathbf{b}$$
$$\mathbf{x} \geq \mathbf{0}$$

where $\mathbf{c}$ is a row vector. We will call this problem the *primal* problem.

Example 5.1.

$$\mathbf{c} = \begin{bmatrix} 2 & 1 \end{bmatrix}$$
$$\mathbf{A} = \begin{bmatrix} 1 & 1 \\ 1 & 0 \end{bmatrix}$$
$$\mathbf{b} = \begin{bmatrix} 2 \\ 1 \end{bmatrix}$$

The program then reads

$$\max 2x_1 + x_2$$
$$x_1 + x_2 \leq 2$$
$$0 \leq x_1 \leq 1$$
$$0 \leq x_2$$

The feasible domain for this example is illustrated (shaded polygon) in Figure 5.1.

Clearly, the solution is on one of the vertices, i.e., either (0,2), (0,0), (1,0), or (1,1), depending on the value of the vector $\mathbf{c}$. The idea of the *simplex algorithm* is to explore the vertices, each time moving to a "better" vertex. A vertex $\mathbf{x}^{(2)}$ is better than a vertex $\mathbf{x}^{(1)}$ if the value of the objective function $\mathbf{cx}^{(2)} > \mathbf{cx}^{(1)}$. How does the algorithm know where to stop? We can apply the first-order conditions that we derived in Chapter 4.

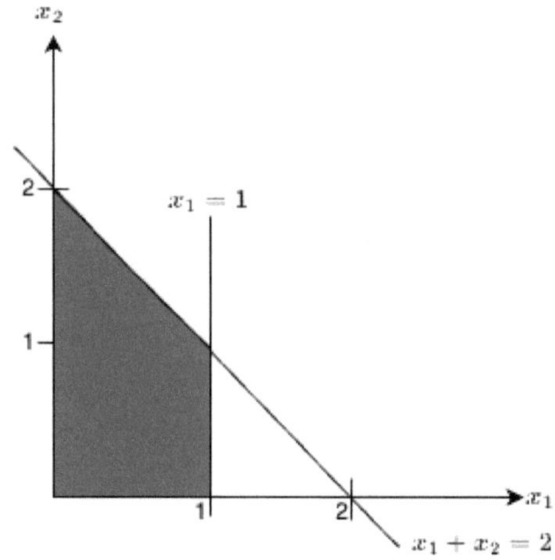

Figure 5.1: Example of linear program. The feasible domain is in gray.

5.1 ▪ Duality

The *dual* problem is defined as

$$\begin{aligned}&\min_{\mathbf{y}} \mathbf{yb}\\ &\mathbf{yA} \geq \mathbf{c}\\ &\mathbf{y} \geq \mathbf{0}\end{aligned}$$

where $\mathbf{y}$ is a row vector.

You will prove in exercise 5.2 that this problem is indeed the Lagrange dual problem of the primal problem. An economic interpretation of this fact is given in the next section.

The *duality* gap is the difference between the optimal solution of the primal and the optimal solution of the primal.

Theorem 5.2 (duality in linear programming). *Suppose that there exists a finite solution* $\mathbf{x}^*$ *to the primal problem. Then there exists a finite solution* $\mathbf{y}^*$ *to the dual problem, and there is no duality gap, i.e.,*

$$\mathbf{cx}^* = \mathbf{y}^*\mathbf{b}$$

5.1.1 ▪ Interpretation of the Dual Problem

This section can be skipped, as it does not pertain to asset pricing or management. We will see, however, in Chapter 23 a link to no-arbitrage theory.

Let $\mathbf{x} \geq \mathbf{0}$ be the vector of goods produced by a company, $\mathbf{c}$ the vector of profit per unit of goods sold, $\mathbf{b}$ the vector of resource capacity, and $\mathbf{A}$ the production matrix. The production constraints can be expressed as

$$\mathbf{Ax} \leq \mathbf{b} \tag{5.1}$$

The problem of a company consists in maximizing the profit $\mathbf{cx}$ under the production constraints. Let $\mathbf{y}$ be the marginal cost of each resource. One could imagine, for instance, that a company rents equipment time from a plant, and that the plant charges a penalty y_m for each (say) hour that equipment m is used beyond the optimal plan. The dual problem consists in minimizing the total marginal cost $\mathbf{yb}$ under the constraint that the marginal cost of each good has to be larger than the profit per good, i.e.,

$$\mathbf{yA} \geq \mathbf{c} \tag{5.2}$$

Indeed, if the marginal cost of each good is lower than the profit of each good, one should produce more since producing more yields additional profit. That is, $\mathbf{y}$ should change until (5.2) is satisfied.

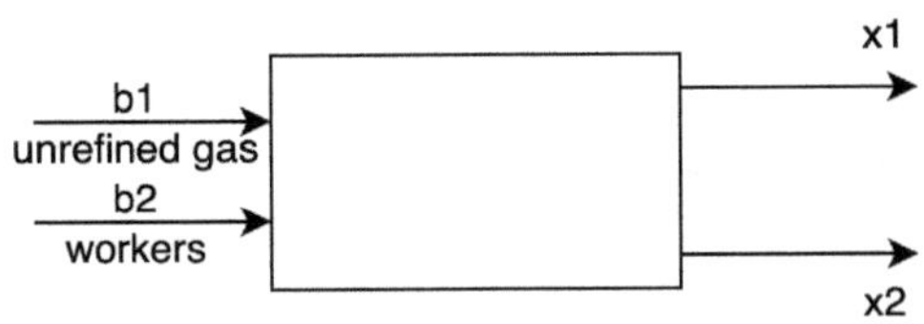

Figure 5.2: Production function: unrefined gas and workers are used to produce gas and artificial diamonds.

Example 5.3. To interpret Figure 5.2, let us consider the example above:

$$\mathbf{A} = \begin{bmatrix} A_{1,1} & A_{1,2} \\ A_{2,1} & A_{2,2} \end{bmatrix} = \begin{bmatrix} 1 & 0 \\ 1 & 1 \end{bmatrix} \tag{5.3}$$

$$\mathbf{b} = \begin{bmatrix} 1 \\ 2 \end{bmatrix} \tag{5.4}$$

$$\mathbf{c} = \begin{bmatrix} 1 & 2 \end{bmatrix} \tag{5.5}$$

There are two goods. Good 1 is gas. Good 2 is artificial diamonds. Let x_1 be gas produced (in kilotons), and let x_2 be diamonds (in kilos). The first resource is storage place in the factory, with capacity $b_1 = 1$ (kiloton), and the second resource is the number of workers, with capacity $b_2 = 2$ (thousands of workers). The entry $A_{m,n}$ represents the amount of resource m needed to produce one unit of good n. The quantity of resource 1 used is clearly

$$A_{1,1}x_1 + A_{1,2}x_2$$

and is limited by b_1. The quantity of resource 2 used is clearly

$$A_{2,1}x_1 + A_{2,2}x_2$$

and is limited by b_2. In our example, it turns out that at the optimum, resources are used at capacity, i.e.,

$$A_{1,1}x_1^* + A_{1,2}x_2^* = b_1$$
$$A_{2,1}x_1^* + A_{2,2}x_2^* = b_2$$

The optimal marginal cost of storage is y_1^* while the optimal marginal cost of workers is y_2^*. The total marginal cost is

$$y_1^* b_1 + y_2^* b_2$$

The marginal cost of (one unit of) good 1 is (at the optimum)

$$y_1^* A_{1,1} + y_2^* A_{2,1}$$

Indeed, we need an amount A_{11} of resource 1 to produce one unit of good 1, and an amount A_{21} of resource 2 to produce one unit of good 1. Likewise, the marginal cost of good 2 is (at the optimum)

$$y_1^* A_{1,2} + y_2^* A_{2,2}$$

In this example, it will turn out that, at the optimum, the marginal cost of each good is equal to its profit, i.e.,

$$\mathbf{y}^* \mathbf{A} = \mathbf{c}$$

In general, however, we will have $\mathbf{y}^* \mathbf{A} \geq \mathbf{c}$.

5.2 ▪ Modeling with Linear Programming

Nonlinear programming (in particular quadratic programming) is often considered to be much more general than linear programming, and one may think that practitioners never use linear programming. In practice, this is not true for *convex programming* problems. What are convex programming problems? They are problems for which every local optimum is global. In other terms, any descent search algorithm, such as the gradient descent algorithm, will converge to the optimum. Many convex optimization problems can be approximated by a linear program.

This section shows two related tricks used to approximate a convex program by a linear program. Interestingly, however, quadratic programs are much easier to handle analytically than linear programs.

Norm Change

For a least squares problem, the goal is to solve

$$\min \mathbf{x}'\mathbf{x}$$

The number $\sqrt{\mathbf{x}'\mathbf{x}}$ is called the L_2 norm of the vector $\mathbf{x}$ and is sometimes denoted by $||\mathbf{x}||_2$. We could be interested in another number, the L_1 norm of the vector $\mathbf{x}$. It is defined as

$$||\mathbf{x}||_1 = \sum |x_i|$$

There is a simple value of minimizing the absolute value. Suppose the problem is

$$\begin{aligned} &\min ||\mathbf{x}||_1 \qquad (5.6) \\ &\mathbf{A}\mathbf{x} = \mathbf{b} \end{aligned}$$

with $\mathbf{x}$ unrestricted, i.e., positive or negative. The trick is to define two positive vectors $\mathbf{x}^+$ and $\mathbf{x}^-$ with the idea that

$$\mathbf{x} = \mathbf{x}^+ - \mathbf{x}^-$$

Thus the original problem (5.6) is equivalent to the problem

$$\begin{aligned} &\min \mathbf{1}'\mathbf{x}^+ + \mathbf{1}'\mathbf{x}^- \qquad (5.7) \\ &\mathbf{A}(\mathbf{x}^+ - \mathbf{x}^-) = \mathbf{b} \\ &\mathbf{x}^+ \geq 0 \qquad (5.8) \\ &\mathbf{x}^- \geq 0 \qquad (5.9) \end{aligned}$$

Approximating Concave Increasing Functions

Figure 5.3 shows how to approximate a concave increasing function $u(c)$ by a piecewise linear utility function $\tilde{u}$. The trick is to first define anchor points $d_0 < d_1 < d_2 < \cdots$. In our example, we set $d_0 = 0$. Then we define three variables y_0, y_1, y_2 such that

$$\begin{aligned} &0 \le y_0 \le d_1 \\ &0 \le y_1 \le d_2 - d_1 \\ &0 \le y_2 \end{aligned}$$

such that

$$c = y_0 + y_1 + y_2$$

The piecewise approximation is then

$$\tilde{u}(y_0, y_1, y_2) = u(d_0) + (u(d_1) - u(d_0))\frac{y_0}{d_1} + (u(d_2) - u(d_1))\frac{y_1}{d_2 - d_1} + u'(d_2)y_2$$

For instance, a program

$$\begin{aligned} &\max_{\mathbf{x},c} u(c) \\ &\mathbf{gx} = c \\ &\mathbf{Ax} = \mathbf{b} \end{aligned}$$

would be approximated by

$$\begin{aligned} &\max_{\mathbf{x},y_0,y_1,y_2} \tilde{u}(y_0, y_1, y_2) \\ &\mathbf{gx} = y_0 + y_1 + y_2 \\ &\mathbf{Ax} = \mathbf{b} \\ &0 \le y_0 \le d_1 \\ &0 \le y_1 \le d_2 - d_1 \\ &0 \le y_2 \end{aligned}$$

Clearly, it is suboptimal to have both $y_0 < d_1$ and y_1 strictly positive. So, if $y_0 < d_1$, we must have $y_1 = 0$.

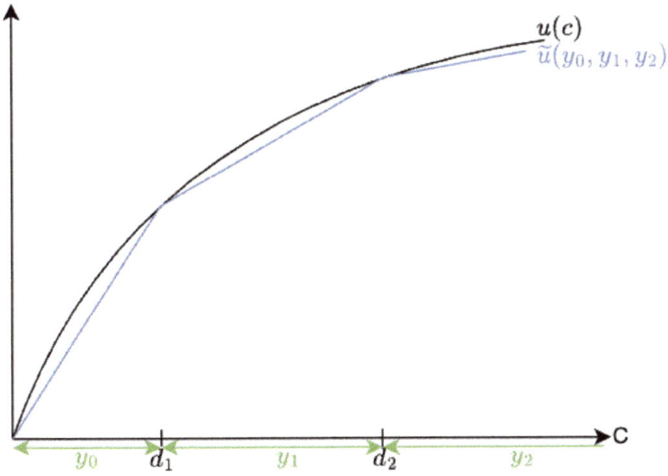

Figure 5.3: Approximating a concave function.

Modeling Transaction Costs

Suppose that when you buy a share of a security you pay $p_0 + c$. If you sell it, you receive $p_0 - c$. The variable $c > 0$ is the transaction cost. To model transaction costs the idea is to decouple the variable x (representing your position at time 1) into two positive variables: the number of shares bought x_b and the number of shares sold x_s at time zero. Thus

$$x = x_b - x_s$$

Now consider a *stochastic program*, where the price $p_1(\omega)$ at time 1 of the stock can take two values, a high value denoted by $p_1(H)$ and a low value denoted by $p_1(L)$, with the same probability. The problem is to maximize the expected value of the utility of wealth. Assuming that the interest rate is zero, it is the following linear program:

$$\begin{aligned}
&\max_{x_b, x_s} \frac{1}{2}u((x_b - x_s)p_1(H)) + \frac{1}{2}u((x_b - x_s)p_1(L)) - x_b(p_0 + c) + x_s(p_0 - c) \\
&x_b \geq 0 \\
&x_s \geq 0
\end{aligned}$$

Because of transaction costs, at the optimum either x_b^* or x_s^* will be zero.

Exercises

5.1. (Medium) The primal and dual problems of linear programming must fall into one of the four logical categories below:

		Primal problem opportunity set	
		Nonempty	Empty
Dual problem opportunity set	Nonempty	1	2
	Empty	3	4

Construct examples for all four cases.

5.2. (Hard) In this problem, we will use the saddle point theorem to prove that if there exists a bounded solution $\mathbf{x}^*$ to the primal, then there is a solution to the $\mathbf{y}^*$ to the dual in linear programming and $\mathbf{cx}^* = \mathbf{y}^*\mathbf{b}$.

(a) Show that the following so-called Lagrangian dual problem is such that if $(\mathbf{x}^*, \mathbf{y}^*)$ solves it, then $\mathbf{x}^*$ solves the primal problem:

$$\min_{\mathbf{y} \geq \mathbf{0}} g(\mathbf{y})$$

$$g(\mathbf{y}) = \max_{\mathbf{x} \geq \mathbf{0}} L(\mathbf{x}, \mathbf{y})$$

where

$$L(\mathbf{x}, \mathbf{y}) = \mathbf{cx} + \mathbf{y}(\mathbf{b} - \mathbf{Ax})$$

(b) Show that the inner maximization is

$$g(\mathbf{y}) = \mathbf{yb} + \max_{\mathbf{x} \geq \mathbf{0}} (\mathbf{c} - \mathbf{yA})\mathbf{x} \tag{5.10}$$

(c) Show that if for some n, $c_n - \sum y_m A_{m,n} > 0$, then the maximization problem in (5.10) is unbounded, and conclude that the constraints to have a bounded dual problem are $\mathbf{c} - \mathbf{yA} \leq \mathbf{0}$.

(d) Use the complementary slackness (of the primal) to prove that the objective function of the dual is $g(\mathbf{y}) = \mathbf{yb}$.

That's it. You have now proved that the dual problem is

$$\begin{aligned} \min_{\mathbf{y} \geq \mathbf{0}} \mathbf{yb} \\ \mathbf{yA} \geq \mathbf{c} \end{aligned}$$

5.3. (Easy) Solve graphically the following linear program:

$$\begin{aligned} \max 2x_1 + 3x_2 \\ x_2 \leq 2 \\ x_1 - x_2 \leq 0 \\ x_1 + x_2 \leq 4 \\ x_1 \geq 0 \\ x_2 \geq 0 \end{aligned}$$

5.4. (Easy) Take $u(x) = \ln(x+1)$. Use $d_1 = 1$ and $d_2 = 2$. Construct $\tilde{u}$ and solve the problem

$$\begin{aligned} \max_{\mathbf{x}, y_0, y_1, y_2} \quad & \tilde{u}(y_0, y_1, y_2) - \gamma x^2 \\ & x = y_0 + y_1 + y_2 \\ & x \leq 1.5 \\ & 0 \leq y_0 \leq d_1 \\ & 0 \leq y_1 \leq d_2 - d_1 \\ & 0 \leq y_2 \end{aligned}$$

for $\gamma = 0$ and $\gamma = 1$.

5.5. (Medium) Solve graphically the primal and dual problem given by (5.3), (5.4), (5.5). Give an economic interpretation of the optimum.

5.6. (Hard) Show that, in our transaction cost problem, the optimal solution satisfies $x_b^* x_s^* = 0$, that is, you either buy or sell.

Chapter 6

Ordinary Least Squares

Notation: In this chapter, we use the traditional notation for regression in financial economics. The parameter vector, instead of $\boldsymbol{\theta}$, is denoted by $\mathbf{b}$. Its true value $\boldsymbol{\theta}^*$ is denoted by $\boldsymbol{\beta}$.

Suppose that a scalar y_t (the *dependent variable* or *regressand*) is related to a K vector of *independent variables* $\mathbf{x}_t$ (or *regressor*) and a *disturbance term*, or *error term*, or *population residual* u_t, according to the *linear regression model*:

$$y_t = \mathbf{x}_t'\boldsymbol{\beta} + u_t \quad t = 1, \ldots, T \tag{6.1}$$

We saw that in ML the dependent variable is called the label, and the independent variable is called features. The quantity $(y_t, \mathbf{x}_t)$ is called a *sample point*, or *observation*, in statistics, and a data point in ML. The dataset (in ML) is formed of T data points and is called a *sample* in statistics with usually

$$K < T$$

The goal is to minimize some measure of the error. In this chapter, we will minimize the sum of squares of the residuals (RSS). The RSS is the square of the L_2 norm of the error. The choice of the L_2 norm is particularly convenient, since this particular estimator of $\boldsymbol{\beta}$ can be calculated purely analytically.[17] Moreover, we will see that the estimator of $\boldsymbol{\beta}$ is efficient under fairly general assumptions and thus is the best we can hope for in terms of generalization error. In the remainder of this book, we will always use the L_2 norm.[18]

Statisticians first investigate the *deterministic case*, which corresponds to the training set in ML. The generalization error is then quantified by assuming that the training set is a sample of a larger population. The population can be modeled in many different ways and referred to as the *stochastic case*. In the first part of this chapter, we analyze the deterministic case. In the second part, we analyze the stochastic case under simple assumptions.

6.1 ▪ The Deterministic Case

The ordinary least squares (OLS) estimate of $\boldsymbol{\beta}$ (denoted by $\hat{\mathbf{b}}$) is the value of $\mathbf{b}$ that minimizes the *residual sum of squares* (RSS)

$$RSS = \sum_{t=1}^{T} (y_t - \mathbf{x}_t'\mathbf{b})^2 \tag{6.2}$$

[17] Some readers may object to the inclusion of such a model in an ML book since this method was known by Gauss, when no "machine" was available. However, most practitioners use linear models at the time of writing this book.

[18] In the chapter on linear programming, we minimize the L_1 norm of the error.

In Exercise 3.5, we saw that the OLS estimate is given by

$$\hat{\mathbf{b}} = \frac{\sum_{t=1}^{T} \mathbf{x}_t y_t}{\sum_{t=1}^{T} \mathbf{x}_t \mathbf{x}_t'} \tag{6.3}$$

It is more convenient to present this model in matrix notation, where we stack every observation vector into a matrix $\mathbf{X}$, which statisticians call the *design matrix*. We will indeed prove (6.3) using the following notation:

$$\mathbf{y} = \begin{bmatrix} y_1 \\ y_2 \\ \vdots \\ y_T \end{bmatrix} \quad \mathbf{X} = \begin{bmatrix} \mathbf{x}_1' \\ \mathbf{x}_2' \\ \vdots \\ \mathbf{x}_T' \end{bmatrix} \quad \mathbf{u} = \begin{bmatrix} u_1 \\ u_2 \\ \vdots \\ u_T \end{bmatrix}$$

We will use the following definitions.

Definition 6.1. *We call the fitted value* $\hat{\mathbf{y}} = \mathbf{X}\hat{\mathbf{b}}$, $\bar{y}$ *the sample average (of the dependent variables), and* $\hat{\mathbf{u}} = \mathbf{y} - \hat{\mathbf{y}}$ *the sample residuals.*

Figure 6.1 illustrates these definitions.

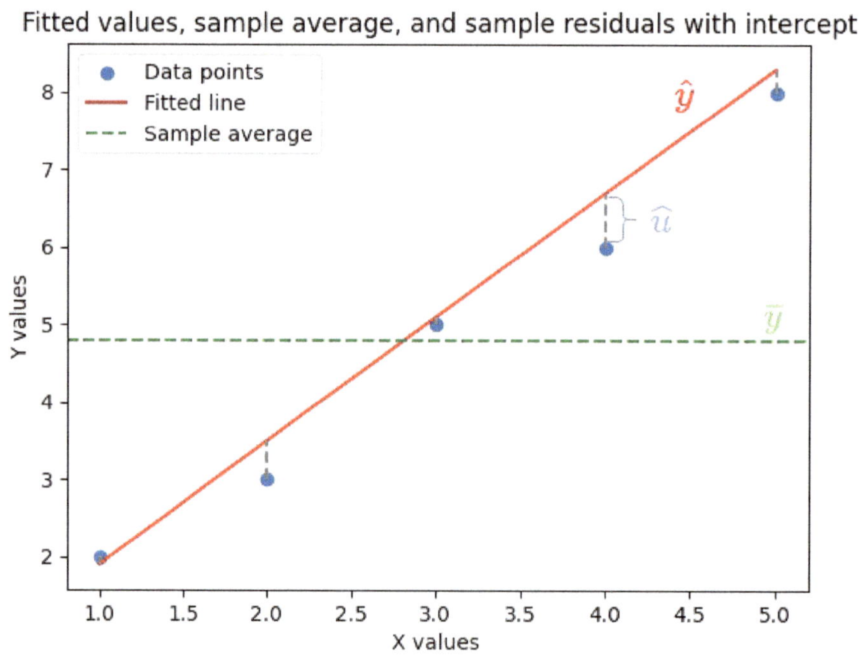

Figure 6.1: Example showing fitted value, sample average, and sample residuals.

The sum of squares of the residuals is thus

$$\begin{aligned} RSS = \mathbf{u}'\mathbf{u} &= (\mathbf{y}' - \mathbf{b}'\mathbf{X}')(\mathbf{y} - \mathbf{X}\mathbf{b}) \\ &= \mathbf{y}'\mathbf{y} - \mathbf{b}'\mathbf{X}'\mathbf{y} - \mathbf{y}'\mathbf{X}\mathbf{b} + \mathbf{b}'\mathbf{X}'\mathbf{X}\mathbf{b} \end{aligned}$$

Since $\mathbf{b}'\mathbf{X}'\mathbf{y} = \mathbf{y}'\mathbf{X}\mathbf{b}$, then

$$RSS = \mathbf{y}'\mathbf{y} - \mathbf{2b}'\mathbf{X}'\mathbf{y} + \mathbf{b}'\mathbf{X}'\mathbf{X}\mathbf{b}$$

Thus the goal is to solve the minimization problem:

$$\min_{\mathbf{b}} f(\mathbf{b}) = \frac{1}{2}\mathbf{b}'\mathbf{X}'\mathbf{X}\mathbf{b} - \mathbf{b}'\mathbf{X}'\mathbf{y} \tag{6.4}$$

We need the following lemma, which you will prove in the exercise.

Lemma 6.2. *For any symmetric matrix* $\mathbf{A}$, $\nabla \mathbf{b}'\mathbf{A}\mathbf{b} = 2\mathbf{A}\mathbf{b}$.

Theorem 6.3. *Suppose* $\mathbf{X}'\mathbf{X}$ *is invertible. The solution to the OLS problem (6.4) is given by*

$$\hat{\mathbf{b}} = (\mathbf{X}'\mathbf{X})^{-1}\mathbf{X}'\mathbf{y} \tag{6.5}$$

This fundamental result (6.5) is called the *normal equations.*

Proof. We calculate

$$\nabla f(\mathbf{b}) = \mathbf{X}'\mathbf{X}\mathbf{b} - \mathbf{X}'\mathbf{y}$$

The first-order necessary conditions for $\hat{\mathbf{b}}$ to be a minimizer are

$$\nabla f(\hat{\mathbf{b}}) = 0$$

from which (6.5) follows if $\mathbf{X}'\mathbf{X}$ is invertible. The Hessian of f is the matrix $\mathbf{X}'\mathbf{X}$, which is strictly positive definite all over the domain. Thus, generalizing Fact 3.4, $\hat{\mathbf{b}}$ is a global minimizer. ■

Example 6.4 (Fama–French model). Suppose that you "regress" a stock return y on the following variables: a constant intercept (1), the market return r^m, the return of small minus big stocks (SMB) r^{SMB}, and the return of high book/market minus low (HML) book/market stocks r^{HML}. We have $T = 5$ observations and $K = 4$ regressors.

$$\mathbf{y} = \begin{bmatrix} \text{return Monday} \\ \text{return Tuesday} \\ \text{return Wednesday} \\ \text{return Thursday} \\ \text{return Friday} \end{bmatrix}$$

$$\mathbf{X} = \begin{bmatrix} 1 & r^m_{\text{M}} & r^{SMB}_{\text{M}} & r^{HML}_{\text{M}} \\ 1 & r^m_{\text{T}} & r^{SMB}_{\text{T}} & r^{HML}_{\text{T}} \\ 1 & r^m_{\text{W}} & r^{SMB}_{\text{W}} & r^{HML}_{\text{W}} \\ 1 & r^m_{\text{Th}} & r^{SMB}_{\text{Th}} & r^{HML}_{\text{Th}} \\ 1 & r^m_{\text{F}} & r^{SMB}_{\text{F}} & r^{HML}_{\text{F}} \end{bmatrix}$$

$$\mathbf{X}' = \begin{bmatrix} 1 & 1 & 1 & 1 & 1 \\ r^m_{\text{M}} & r^m_{\text{T}} & r^m_{\text{W}} & r^m_{\text{Th}} & r^m_{\text{F}} \\ r^{SMB}_{\text{M}} & r^{SMB}_{\text{T}} & r^{SMB}_{\text{W}} & r^{SMB}_{\text{Th}} & r^{SMB}_{\text{F}} \\ r^{HML}_{\text{M}} & r^{HML}_{\text{T}} & r^{HML}_{\text{W}} & r^{HML}_{\text{Th}} & r^{HML}_{\text{F}} \end{bmatrix}$$

Then

$$\mathbf{X}'\mathbf{X} =$$

$$\begin{bmatrix} 5 & \sum_{t=1}^{5} r_t^m & \sum_{t=1}^{5} r_t^{SMB} & \sum_{t=1}^{5} r_t^{HML} \\ \sum_{t=1}^{5} r_t^m & \sum_{t=1}^{5} (r_t^m)^2 & \sum_{t=1}^{5} (r_t^m)(r_t^{SMB}) & \sum_{t=1}^{5} (r_t^m)(r_t^{HML}) \\ \sum_{t=1}^{5} r_t^{SMB} & \sum_{t=1}^{5} (r_t^m)(r_t^{SMB}) & \sum_{t=1}^{5} (r_t^{SMB})^2 & \sum_{t=1}^{5} (r_t^{SMB})(r_t^{HML}) \\ \sum_{t=1}^{5} r_t^{HML} & \sum_{t=1}^{5} (r_t^m)(r_t^{HML}) & \sum_{t=1}^{5} (r_t^{SMB})(r_t^{HML}) & \sum_{t=1}^{5} (r_t^{HML})^2 \end{bmatrix}$$

and

$$\mathbf{X}'\mathbf{y} = \begin{bmatrix} \sum_{t=1}^{5} r_t \\ \sum_{t=1}^{5} r_t^m r_t \\ \sum_{t=1}^{5} r_t^{SMB} r_t \\ \sum_{t=1}^{5} r_t^{HML} r_t \end{bmatrix}$$

The next section can be skipped at first reading but can be seen as a mathematical "preparation" for Chapter 23 on the stochastic discount factor.

6.1.1 ▪ Approximation versus Interpolation

There is a unified way of looking at underdetermined, exactly determined, and overdetermined systems of linear equations. All three can be solved by premultiplying the right-hand side $\mathbf{b}$ by a pseudo-inverse matrix.

When $K < T$, typical of linear regressions, we have the *approximation* problem $\mathbf{Xb} \approx \mathbf{y}$. Since the system of equations $\mathbf{Xb} = \mathbf{y}$ is overdetermined, the RSS cannot be equal to zero. As we saw, the OLS solution consists in taking for the pseudo-inverse the left Moore–Penrose inverse $(\mathbf{XX}')^{-1}\mathbf{X}'$.

When $K = T$, the $\mathbf{X}$ matrix is square and (usually) can be inverted, so that the pseudo-inverse is the inverse $\mathbf{X}^{-1}$.

When $K > T$ there are several solutions to the system $\mathbf{y} = \mathbf{Xb}$, which is underdetermined. This corresponds to different ways of *interpolating* the points. A solution that is popular[19] is to take for the pseudo-inverse the right Moore–Penrose inverse $\mathbf{X}^{'}(\mathbf{XX}')^{-1}$.

We summarize these important facts in Table 6.1.[20] One can easily remember all solutions by looking at the dimension of the $\mathbf{X}$ matrix.

Problem	System	Solution $\mathbf{b} =$
Approximation	$\begin{bmatrix} \\ \mathbf{X} \\ \\ \end{bmatrix} \begin{bmatrix} \mathbf{b} \end{bmatrix} \approx \begin{bmatrix} \\ \mathbf{y} \\ \\ \end{bmatrix}$	$(\mathbf{XX}')^{-1}\mathbf{X}^{'}\mathbf{y}$
Exact	$\begin{bmatrix} \mathbf{X} \end{bmatrix} \begin{bmatrix} \mathbf{b} \end{bmatrix} = \begin{bmatrix} \mathbf{y} \end{bmatrix}$	$\mathbf{X}^{-1}\mathbf{y}$
Interpolation	$\begin{bmatrix} \quad \mathbf{X} \quad \end{bmatrix} \begin{bmatrix} \\ \mathbf{b} \\ \\ \end{bmatrix} = \begin{bmatrix} \mathbf{y} \end{bmatrix}$	$\mathbf{X}^{'}(\mathbf{XX}')^{-1}\mathbf{y}$

Table 6.1: Inverses and pseudo-inverses.

[19] You saw in Exercise 3.10 that the solution has minimal length.

[20] We assume that the inverses exist.

6.1.2 ▪ Properties

The OLS sample residuals are given by

$$\begin{aligned}\hat{\mathbf{u}} &= \mathbf{y} - \mathbf{X}\hat{\mathbf{b}} \\ &= \mathbf{y} - \mathbf{X}(\mathbf{X}'\mathbf{X})^{-1}\mathbf{X}'\mathbf{y} \\ &= (\mathbf{I} - \mathbf{X}(\mathbf{X}'\mathbf{X})^{-1}\mathbf{X}')\mathbf{y}\end{aligned}$$

Define the projector matrix:

$$\mathbf{M}_X = \mathbf{I} - \mathbf{X}(\mathbf{X}'\mathbf{X})^{-1}\mathbf{X}'$$

It is easy to see that $\mathbf{M}_X$ is symmetric,

$$\mathbf{M}_X = \mathbf{M}_X'$$

idempotent,

$$\mathbf{M}_X\mathbf{M}_X = \mathbf{M}_X$$

and orthogonal to the columns of $\mathbf{X}$,

$$\mathbf{M}_X\mathbf{X} = \mathbf{0}$$

Thus the OLS sample residuals are orthogonal to the explanatory variables:

$$\hat{\mathbf{u}}'\mathbf{X} = \mathbf{y}'\mathbf{M}_X\mathbf{X} = \mathbf{0}'$$

Figure 6.2 illustrates orthogonality.

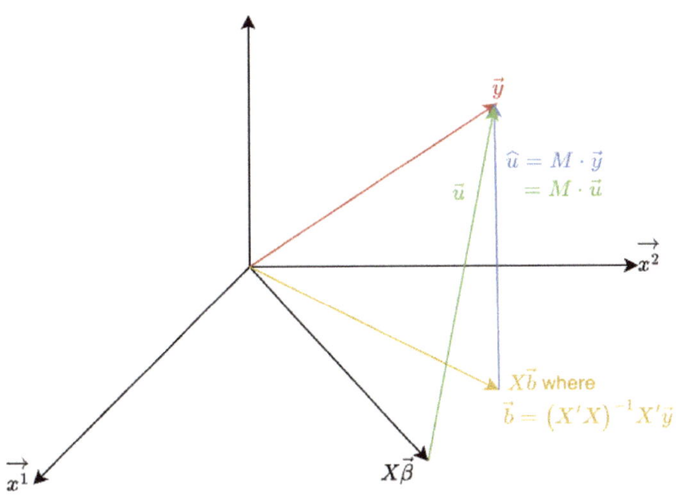

Figure 6.2: The sample residuals are orthogonal with the projection on the space of the independent variables.

The OLS sample residuals $\hat{\mathbf{u}}$ should be distinguished from the *population residuals* $\mathbf{u}$. The sample residual is constructed from the sample estimate $\hat{\mathbf{b}}$ of $\boldsymbol{\beta}$, whereas the population residual is a hypothetical construct based on the true population value $\boldsymbol{\beta}$ from (6.1). The relation between the two is given by

$$\hat{\mathbf{u}} = \mathbf{M}_X\mathbf{y} = \mathbf{M}_X(\mathbf{X}\boldsymbol{\beta} + \mathbf{u}) = \mathbf{M}_X\mathbf{u}$$

Likewise, the difference between the OLS estimate $\hat{\mathbf{b}}$ and the true population parameter $\boldsymbol{\beta}$ is given by

$$\begin{aligned}\hat{\mathbf{b}} &= (\mathbf{X}'\mathbf{X})^{-1}\mathbf{X}'\mathbf{y} \\ &= (\mathbf{X}'\mathbf{X})^{-1}\mathbf{X}'(\mathbf{X}\boldsymbol{\beta}+\mathbf{u}) = \boldsymbol{\beta} + (\mathbf{X}'\mathbf{X})^{-1}\mathbf{X}'\mathbf{u}\end{aligned} \tag{6.6}$$

The quality of the fit in an OLS regression is sometimes described in terms of the sample multiple correlation coefficient, also called **coefficient of determination**, or R^2. The *uncentered* R^2, denoted by R_u^2, is defined as the sum of squares of the fitted values of the regression as a fraction of the sum of squares of the dependent variables:

$$R_u^2 = \frac{\hat{\mathbf{y}}'\hat{\mathbf{y}}}{\mathbf{y}'\mathbf{y}}$$

It is often better to compare the error from linear regression to the error we would make by taking as our *predictor* of y_t the sample average $\bar{y}$ of the dependent variables. We can thus define the *explained sum of squares* and the *total sum of squares* as

$$\begin{aligned}\text{explained SS} &= \|\hat{\mathbf{y}}-\bar{y}\mathbf{1}\|^2 \\ \text{total SS} &= \|\mathbf{y}-\bar{y}\mathbf{1}\|^2\end{aligned}$$

The *centered* R^2, denoted by R_c^2, is the ratio

$$R_c^2 = \frac{\text{explained SS}}{\text{total SS}}$$

When the $\mathbf{X}$ matrix has a column of ones, it can be proved that

$$\text{total SS} = \text{explained SS} + \text{RSS} \tag{6.7}$$

This translates into the following result for the centered R-squared:

$$R_c^2 = 1 - \frac{\|\hat{\mathbf{u}}\|^2}{\|\mathbf{y}-\bar{y}\mathbf{1}\|^2}$$

Since a linear regression provides a better estimator $\hat{\mathbf{y}}$ than just the average of the observations, the residual $\|\hat{\mathbf{u}}\|^2$ is smaller than $\|\mathbf{y}-\bar{y}\mathbf{1}\|^2$, and thus R_c^2 is a value between 0 and 1. The closer to 1, the better your fit.

6.2 ▪ The Stochastic Case

The advantage of embedding our regression problem into a stochastic problem is that we can quantify generalization errors.

Apart from the parameter $\boldsymbol{\beta}$, all variables in (6.1) are random. We can rewrite (6.1) as

$$y_t(\omega) = \mathbf{x}_t'(\omega)\boldsymbol{\beta} + u_t(\omega) \tag{6.8}$$

Equation (6.8) can be understood intuitively in two different ways. The first interpretation is that the chance parameter ω identifies which observations are taken in the sample. For example, a sample corresponding to data taken at $T = 10$ different times is split into two sample points, ω_1 and ω_2. The first sample point, ω_1, is made of observations at time $t \in \{1,2,3,4,5\}$, while the second sample point is made of observations taken at time $t \in \{6,7,8,9,10\}$. Since in ML one takes all the data in the training set, we must resort to a second interpretation for statistical analysis. At each time, nature chooses a particular outcome out of an (unknown) sample space. This is the interpretation we will use in this book. Of course, when one looks at past observations, this viewpoint is nonintuitive, since there is (barring measurement errors) only one history of, e.g., stock prices.

Definition 6.5. *An estimator $\hat{\boldsymbol{\theta}}(\omega)$ of $\boldsymbol{\theta}^*$ is*

- *unbiased if $E[\hat{\boldsymbol{\theta}}] = \boldsymbol{\theta}^*$*
- *efficient if the variance-covariance matrix of any other estimator $\tilde{\boldsymbol{\theta}}(\omega)$ of $\boldsymbol{\theta}^*$ differs from the variance-covariance matrix of $\hat{\boldsymbol{\theta}}(\omega)$ by a positive definite matrix*
- *consistent if, when taking an infinite number of sample points, $\hat{\boldsymbol{\theta}}(\omega) \to \boldsymbol{\theta}^*$.*

When $\theta(\omega)$ is a scalar, efficiency means that $\tilde{\theta}(\omega)$ has a higher variance than $\hat{\theta}(\omega)$.

Different statistical assumptions lead to different results. We are not focusing in this book on the distributional properties of our estimators. The following classical (distribution-free) result is a version of the **Gauss–Markov theorem**. Since most of our estimation problems will be about forecasting, we define linear estimators in this context.

Definition 6.6. *An estimator $\hat{y}$ of y_{t+1} is called linear if it is a linear function of the previous values of the dependent variables up to time t.*

Note that the coefficients of the forecast can be functions of time.

Theorem 6.7. *Suppose* (a) $\mathbf{x}_t$ *is deterministic and* (b) $u_t(\omega)$ *is IID with mean* 0 *and variance* σ^2. *Then $\hat{\mathbf{b}}(\omega)$ given by (6.3) is efficient in the class of linear estimators.*

This is called the **BLUE** property of OLS: the OLS estimator is the Best Estimator in the class of Linear Estimators, and it is Unbiased.

What we will be interested in (in this book) is the following assumption.

Assumption 1. (a) $\mathbf{x}_t(\omega)$ stochastic (b) conditional on the full matrix $\mathbf{X}$, the vector $\mathbf{u}(\omega)$ has mean zero with variance $\sigma^2\mathbf{V}$.

When $\mathbf{V}$ is diagonal, we say that the errors exhibit *heteroskedasticity*. Otherwise, they are *autocorrelated.*

From (6.6), we see that

$$\hat{\mathbf{b}}(\omega) - \boldsymbol{\beta} = (\mathbf{X}'(\omega)\mathbf{X}(\omega))^{-1}\mathbf{X}'(\omega)\mathbf{u}(\omega)$$

Taking conditional expectations,

$$E[\hat{\mathbf{b}} - \boldsymbol{\beta}|\mathbf{X}] = (\mathbf{X}'\mathbf{X})^{-1}\mathbf{X}'E[\mathbf{u}] = \mathbf{0}$$

Thus the OLS estimator $\hat{\mathbf{b}}(\omega)$ is unbiased:

$$E[\hat{\mathbf{b}} - \boldsymbol{\beta}] = E[E[\hat{\mathbf{b}} - \boldsymbol{\beta}|\mathbf{X}]] = \mathbf{0}$$

We can calculate its conditional variance by

$$\begin{aligned} Var[\hat{\mathbf{b}} - \boldsymbol{\beta}|\mathbf{X}] &= \mathbf{E}[(\hat{\mathbf{b}} - \boldsymbol{\beta})(\hat{\mathbf{b}} - \boldsymbol{\beta})'|\mathbf{X}] \\ &= E[((\mathbf{X}'\mathbf{X})^{-1}\mathbf{X}'\mathbf{u})((\mathbf{X}'\mathbf{X})^{-1}\mathbf{X}'\mathbf{u})'|\mathbf{X}] \\ &= E[(\mathbf{X}'\mathbf{X})^{-1}\mathbf{X}'\mathbf{u}'\mathbf{u}\mathbf{X}(\mathbf{X}'\mathbf{X})^{-1})|\mathbf{X}] \\ &= \sigma^2(\mathbf{X}'\mathbf{X})^{-1}\mathbf{X}'\mathbf{V}\mathbf{X}(\mathbf{X}'\mathbf{X})^{-1} \end{aligned}$$

Theorem 6.7 applies only to the case where $\mathbf{V}$ is the identity matrix. If it is not, then an efficient estimator is given by generalized least squares (GLS), which we will see in Chapter 7.

6.2.1 ▪ Statistical Tests

Statisticians typically use the law of large numbers to estimate the expected value of an estimator. To estimate its distribution, the central limit theorem is the tool of choice. Every statistical package returns the R^2 as well as t-statistics and, sometimes, F-statistics of linear regression; see Example 6.8 below. Although many of these measures do not easily generalize to the nonlinear case, it is important to grasp what they mean. These tests help us identify spurious regressions.

T-test

We can easily calculate an estimator $\hat{\sigma}^2$ of σ^2 in the heteroskedastic case, i.e., in the case where $\mathbf{V} = \mathbf{I}$,

$$\hat{\sigma}^2 = \frac{\hat{\mathbf{u}}'\hat{\mathbf{u}}}{T - K}$$

It can be proved that it is unbiased, i.e., $E[\hat{\sigma}^2] = \sigma^2$. To test that β_i is significantly different from zero, one defines the *t-statistic*:

$$t_i(\omega) = \frac{\hat{b_i}(\omega)}{s_i(\omega)}$$

where $s_i = \hat{\sigma}((\mathbf{X}'\mathbf{X})^{-1})_{i,i}$. If the errors are normally distributed, then $t_i(\omega)$ has the Student's t distribution, with $T - K$ degrees of freedom. For instance, with 1 degree of freedom

$$P(t_i < 6.313) = 95\%$$

while with 10 degrees of freedom

$$P(t_i < 1.812) = 95\%$$

In other words, if the observed value $t_i(\omega) > 1.812$, there is a 95% chance that β_i is larger than zero. Thus the ith regressand is significant and should be included in the regression. A popular method to get rid of insignificant regressands is to try all possible subsets of regressands and get rid of those with the smallest t values. A more economical method is LASSO.

F-test

While the T-test allows us to accept or reject the hypothesis that a particular parameter takes a particular value, the F-test (or Wald test) quantifies whether joint linear restrictions on the values of the parameters hold. The corresponding null hypothesis is, for given matrix $\mathbf{R}$ and vector $\mathbf{r}$,

$$H_0 : \mathbf{R}\boldsymbol{\beta} = \mathbf{r}$$

The interested reader will learn more about this test in standard statistics books. We will not make use of this test in this book.

Example 6.8. In this example, we ran an OLS for the Fama–French three-factor model from 2013-04-01 to 2023-03-31. The dependent variable is excess market return, and the independent variables are excess SMB (small minus big) and excess HML (high minus low). The results are shown in the following table:

	Estimate	Std. Error	t-statistic
const	1.0313	0.395	2.612
Excess_HML	0.0133	0.109	0.123
R-squared	0.076		
Adjusted R-squared	0.061		
F-statistic	4.840		
Prob(F-statistic)	0.00956		

Definition 6.9. *The p-value is defined as the probability, under the null hypothesis H_0, of observing a statistic at least as extreme as the one calculated from the sample data. If T is our test statistic and t is the observed value, the p-value can be expressed as*

$$p\text{-}value = P(T \geq t | H_0)$$

A common threshold for statistical significance is α, often set at 0.05. If the p-value $\leq \alpha$, then the result is considered statistically significant, and we reject H_0. Otherwise, we fail to reject H_0.

1. The t-statistic for *Excess_SMB* is 3.098, and its p-value is less than 0.05. This suggests that the *Excess_SMB* is statistically significant in explaining the *Excess_Market Returns*.

2. The t-statistic for *Excess_HML* is 0.123, and its p-value is 0.903. It suggests that *Excess_HML* is not statistically significant in explaining the *Excess_Market Returns* at the 5% significance level.

The F-statistic for the model is 4.840, and the p-value is 0.00956. This suggests that the overall regression model is statistically significant, and the combination of *Excess_SMB* and *Excess_HML* can significantly explain the variation in *Excess_Market Returns*.

6.3 ▪ Radial Basis Function Interpolation

A problem with the specification (6.1) is that the model (i.e., the function f) is **linear in both the input data and the parameter**. One can obtain extra mileage in OLS if one enlarges the space of independent variables by adding nonlinear functions of the original independent variables. Typical examples of one-dimensional models are expansions in *basis functions*:

- polynomials: $f(x; \boldsymbol{\theta}) = \theta_0 + \theta_1 x + \theta_2 x^2 + \cdots$
- cosine Fourier series: $f(x; \boldsymbol{\theta}) = \theta_0 + \theta_1 \cos(\frac{x}{\max\{x_t\}}) + \theta_2 \cos(\frac{2x}{\max\{x_t\}}) + \cdots$
- splines
- kernels.

In one dimension, the problem has been well studied and is completely understood. When the number of dimensions is large, however, the number of basis functions grows exponentially. A better idea is to do radial basis function interpolation. In that specification,

$$f(\mathbf{x};\boldsymbol{\theta}) = \sum_{t=1}^{T} \theta_t \varphi(||\mathbf{x} - \mathbf{x}_t||)$$

For instance, a Gaussian basis function would be $\varphi(x) = \exp(-x^2)$. This specification is particularly useful when the observations $\{\mathbf{x}_t\}$ do not fall on a grid, which is almost always the case in finance.

We would then take $K = T$ (number of parameters = number of observations) and solve the problem by either the gradient method or the normal equations.

We will see in Chapter 11 a different architecture, namely, feedforward neural networks, which results not only in better computational performance in high dimensions but genuine nonlinearity in the parameter $\boldsymbol{\theta}$. The following example shows why nonlinearity in the parameter is important.

Example 6.10. Suppose that the true model is

$$f^*(\mathbf{x}) = \exp(-(x_1^2 - x_2^2)) \tag{6.9}$$

A possible specification, linear in the parameters, is

$$f_1(\mathbf{x},\boldsymbol{\theta}) = \theta_1 + \theta_2 x_1 + \theta_3 x_2 + \theta_4 x_1^2 + \theta_5 x_2^2 + \theta_6 x_1 x_2 + \cdots \tag{6.10}$$

while a possible alternate specification, nonlinear in the parameters, would be

$$f_2(\mathbf{x},\boldsymbol{\theta}) = g(\theta_1 + \theta_2 x_1 + \theta_3 x_2 + \theta_4 x_1^2 + \theta_5 x_2^2 + \theta_6 x_1 x_2 + \cdots) \tag{6.11}$$

If one is lucky enough to identify a correct approximation of the function g, i.e., in this example, the exponential, then clearly f_2 is an easier model to fit than f_1. In practice, we will let an algorithm figure out the function g itself (see Exercise 6.6).[21]

Exercises

6.1. (Hard) When the $\mathbf{X}$ matrix has a column of ones, prove (6.7), i.e.,

total SS = explained SS + RSS

6.2. (Medium) Answer yes or no to the following questions. If you answer no, provide a counterexample:
(1) Is an unbiased estimator necessarily consistent?
(2) Is a consistent estimator necessarily unbiased?
(3) Is an efficient estimator necessarily unbiased?
(4) Is an unbiased estimator necessarily efficient?

[21] If g is known a priori, then transforming the data, i.e., setting $g^{-1}(y_t)$ as the dependent variable, transforms the problem into a linear problem. In high dimensions, it is difficult for the experimenter to do it effectively, and one resorts to ML.

6.3. (Easy) Prove Lemma 6.2, i.e., that for any symmetric matrix $\mathbf{A}$, $\nabla \mathbf{b}'\mathbf{A}\mathbf{b} = 2\mathbf{A}\mathbf{b}$.

Hint: Observe that, for any l,

$$\sum_{1\le i\le K}\sum_{1\le j\le K} b_i b_j A_{i,j} = b_l^2 A_{l,l} + \sum_{1\le i<j\le K} 1\{i \ge l\} 2 b_i b_j A_{i,j} + \sum_{1\le i<j\le K} 1\{i < l\} 2 b_i b_j A_{i,j}$$

and then differentiate with respect to b_l.

6.4. (Medium) Prove the Gauss–Markov theorem (Theorem 6.7).

6.5. Use the software of your choice to do a linear regression by OLS of the Fama–French model for monthly returns during the period January 2017 to January 2021. The dependent variable y is the return on the portfolio SMALLloBM and can be found in the worksheet "Returns."[22] The independent variables can be found in the worksheet "Factors."

6.6. (Medium) Generate some data from the example, i.e., equation (6.9). For the function g in the second model (6.11), choose

$$g(z) = \sigma_1 + \sigma_2 z + \sigma_3 z^2$$

and content yourself with a quadratic approximation for z, i.e., $\theta_k = 0$ for all $k > 6$. The result will be a nonlinear model of degree 4, with a set of constraints:

$$\begin{aligned} z_1 &= \theta_1 + \theta_2 x_{1,1} + \theta_3 x_{2,1} \\ &\dots \\ z_T &= \theta_1 + \theta_2 x_{1,T} + \theta_3 x_{2,T} \end{aligned}$$

Model 2 has order 4 and nine parameters. The first model (6.10) needs 15 parameters to be a completely general model of order 4. Calibrate both models, and check which one gives you better results. Use the `fmincon` function (optimization under constraints) in MATLAB, or the Solver in Excel to calibrate model 2.

[22] See supplementary material at https://bookstore.siam.org/ot195/bonus.

Chapter 7

Generalized Least Squares (GLS)

We consider the linear regression problem

$$\mathbf{y}(\omega) = \mathbf{X}(\omega)\boldsymbol{\beta} + \mathbf{u}(\omega) \tag{7.1}$$

We are going to reuse Assumption 1 from Chapter 6, that is,

$$Var[\mathbf{u}|\mathbf{X}] = \sigma^2\mathbf{V}$$

But now we suppose that the errors $\mathbf{u}(\omega)$ are autocorrelated, that is, $\mathbf{V}$ is no longer diagonal. In this case, the Gauss–Markov theorem does not hold anymore. A simple transformation of variables allows us to return to the case where the Gauss–Markov theorem applies. We decompose $\mathbf{V}$ into

$$\mathbf{V}^{-1} = \mathbf{L}'\mathbf{L}$$

Now define the transformed variables

$$\begin{aligned}\tilde{\mathbf{u}} &= \mathbf{L}\mathbf{u}\\ \tilde{\mathbf{X}} &= \mathbf{L}\mathbf{X}\\ \tilde{\mathbf{y}} &= \mathbf{L}\mathbf{y}\end{aligned}$$

Thus, our regression

$$\mathbf{y}(\omega) = \mathbf{X}(\omega)\boldsymbol{\beta} + \mathbf{u}(\omega) \tag{7.2}$$

can be rewritten

$$\tilde{\mathbf{y}}(\omega) = \tilde{\mathbf{X}}(\omega)\boldsymbol{\beta} + \tilde{\mathbf{u}}(\omega) \tag{7.3}$$

Clearly,

$$E[\tilde{\mathbf{u}}\tilde{\mathbf{u}}'|\mathbf{X}] = \sigma^2\mathbf{L}(\mathbf{L}'\mathbf{L})^{-1}\mathbf{L}' = \sigma^2\mathbf{I}$$

We are thus in the situation of the Gauss–Markov theorem. The GLS estimator of $\boldsymbol{\beta}$, given by

$$\begin{aligned}\tilde{\mathbf{b}} &= (\tilde{\mathbf{X}}'\tilde{\mathbf{X}})^{-1}\tilde{\mathbf{X}}'\tilde{\mathbf{y}}\\ &= (\mathbf{X}'\mathbf{V}^{-1}\mathbf{X})^{-1}\mathbf{X}'\mathbf{V}\mathbf{y}\end{aligned}$$

is efficient (conditional on $\mathbf{X}$). In other terms, conditional on $\mathbf{X}$, it is the minimum variance estimator (among the class of linear unbiased estimators).

Example 7.1 (AR(1) process). Consider a special case of the AR(1) process. We will see the general case in Chapter 12 on long short-term memory networks:

$$u_t(\omega) = \rho u_{t-1}(\omega) + \varepsilon_t(\omega) \tag{7.4}$$

where ε_t is IID (not necessarily Gaussian) with mean zero and variance σ^2. You will show in Exercise 7.1 that

$$\mathbf{V} = \frac{\sigma^2}{1-\rho^2}\begin{bmatrix} 1 & \rho & \dots & \rho^{T-1} \\ \rho & \rho^2 & \dots & \rho^{T-2} \\ \vdots & \vdots & \dots & \vdots \\ \rho^{T-1} & \rho^{T-2} & \dots & 1 \end{bmatrix} \tag{7.5}$$

We have a chicken and egg problem: In order to estimate $\boldsymbol{\beta}$ one must know $\mathbf{V}$. However, in order to estimate $\mathbf{V}$, one must know $u_t(\omega)$, and thus know $\boldsymbol{\beta}$. The next section discusses this problem.

7.1 ▪ Estimation of the Variance Matrix of Residuals from the Data

A solution is to do a joint estimation of the parameters $(\boldsymbol{\beta}, \mathbf{V})$. For the AR(1) process, when the white noise ε_t is Gaussian, this is a well-known problem that can be solved by maximum likelihood (see Exercise 3.4).

In other cases, we can use an iterative algorithm to solve the problem: start with an initial value of $\mathbf{V}$ at iteration zero, labeled $\mathbf{V}^{(0)}$, then calculate a value of $\tilde{\mathbf{b}}$ by GLS at iteration 1, labeled $\tilde{\mathbf{b}}^{(1)}$, then use it to calculate residuals $\hat{\mathbf{u}}^{(1)}$ at iteration 1, from which we can estimate a matrix $\mathbf{V}^{(1)}$ and then continue. This is called the iterated Cochrane–Orcutt method.

ALGORITHM 7.1. Cochrane–Orcutt method.

$\mathbf{V}^{(0)} = \mathbf{I}$
$k = 1$
`Repeat`
  `Calculate` $\tilde{\mathbf{b}}^{(k)} = (\mathbf{X}'(\mathbf{V}^{(k-1)})^{-1}\mathbf{X})^{-1}\mathbf{X}'\mathbf{V}^{(k-1)}\mathbf{y}$
  `Set` $\mathbf{L}^{(k-1)} = (\mathbf{V}^{(k-1)})^{-1/2}$ `by Cholesky decomposition`
  `Set` $\hat{\mathbf{u}}^{(k)} = \mathbf{y} - \mathbf{L}^{(k-1)}\mathbf{X}\mathbf{b}^{(k)}$
  `Set` $\mathbf{V}^{(k)} = (\,\hat{\mathbf{u}}^{(k)})'\,\hat{\mathbf{u}}^{(k)}$
  `Set` $k = k+1$
`Until converge`

In our AR(1) example, this translates more simply into calculating at each iteration k an estimate of ρ given by

$$\hat{\rho}^{(k)} = \frac{\sum_{t=1}^{T} \hat{u}_{t-1}^{(k-1)}\hat{u}_t^{(k-1)}}{\sum_{t=1}^{T}(\hat{u}_t^{(k-1)})^2}$$

Under some conditions, the Cochrane–Orcutt method will converge; see Hamilton [19, p. 225]. We will return to this algorithm in Chapter 8 on the generalized method of moments, and in Chapter 13 on generative adversarial networks. Both are essentially the same method.

Exercises

7.1. (Medium) Show that the variance-covariance matrix of (7.4) is given by $\mathbf{V}$. Calculate $\mathbf{L}$ by Cholesky decomposition.

7.2. Estimate the parameters of the Fama–French model from Exercise 6.5. Assume model (7.4). Vary ρ until you obtain the best RSS, and report the results.

7.3. Take $T = 10$ and the following data:

t	1	2	3	4	5	6	7	8	9	10
x_t	1.5	3	3.7	3.9	2.1	2.0	1.9	1.6	1.5	1.4

Take $\beta = 1$. Vary $\rho \in \{0, 0.4, 0.9\}$ and $\sigma \in \{0.1, 0.5, 1\}$. For all nine combinations of these values, generate data (by Monte Carlo) from the following model (with v_t a uniform variate)

$$u_t = \rho u_{t-1} + \sigma^2 \left(v_t - \frac{1}{2} \right)$$
$$y_t = \beta x_t + u_t$$

and plot $b^{(k)}$ as well as the RSS in the Cochrane–Orcutt algorithm as a function of the iteration k. Take 10 iterations.

7.4. (Easy) Assume $T = 5$ and

$$\mathbf{V} = \begin{bmatrix} 1 & 0 & 0 & 0 & 0 \\ 0 & 2.2 & 0 & 0 & 0 \\ 0 & 0 & 0.3 & 0 & 0 \\ 0 & 0 & 0 & 0.7 & 0 \\ 0 & 0 & 0 & 0 & 0.6 \end{bmatrix}$$

What is the formula for the GLS estimator of $\boldsymbol{\beta}$ as a function of $\mathbf{X}$ and $\mathbf{y}$?

Chapter 8

The Generalized Method of Moments (GMM)

The goal of the generalized method of moments (GMM) is to find an approximation $\hat{\boldsymbol{\theta}}$ of a parameter $\boldsymbol{\theta}^*$ that satisfies

$$E[\mathbf{p}(\mathbf{z}, \boldsymbol{\theta}^*)] = \mathbf{0} \tag{8.1}$$

where $\mathbf{z}$ is a vector of random variables from which the data is sampled. It can be a time series, a cross-section, or both. The vector $\hat{\boldsymbol{\theta}}$ is K-dimensional. The function $\mathbf{p}$ is some function of the data and parameters. This is a very general formulation. In this book, we will apply it to

- models of return
- stochastic discount factor models.

GMM can be applied to overidentified models or to exactly identified models. For exactly identified models, $\mathbf{p}$ is typically the model error itself (or its square). For overidentified models, $\mathbf{p}$ would be (in this book) the gradient of some weighted squares of the errors so that the number of equations (called moment conditions) is the same as the number of parameters.

On the theoretical standpoint, GMM has some advantages that the basic OLS theory we saw does not have. It handles *stationary* series, i.e., series for which the covariance of the residuals, namely, $E[u_t u_{t+j}]$, depends only on j. It is thus more general than an IID sequence, for which the latter is equal to zero.[23]

We start the chapter with elementary examples of each model type, exactly identified, and then overidentified.

8.1 ▪ Example of an Exactly Identified Model

GMM generalizes the method of moments, familiar from introductory statistics. This simple example is meant only to introduce notation and will not bring the reader anything new.

Suppose that we take T observations $(z_1, \ldots, z_T)$ of the scalar random variable $z(\omega)$. Then an estimator $\hat{\theta}$ of $\theta^* \equiv E[z]$ is given by the sample moment

$$\hat{\theta} = \frac{1}{T}\sum_{t=1}^{T} z_t \tag{8.2}$$

[23]More exactly, the original theory handles strictly stationary series.

It is convenient to use the same notation for the random variable and for its observations and to have a special notation for the sample average. To this effect, Hansen [20] introduced the operator E_T for the sample average. We can thus rewrite (8.2) as

$$\hat{\theta} = E_T[z] \equiv \frac{1}{T} \sum_{t=1}^{T} z_t$$

The function

$$p(z, \theta^*) = z - \theta^*$$

is the model error. The relation (8.1) $E[p(z, \theta^*)] = 0$ drives the expected error to zero, namely,

$$E[z] - \theta^* = 0 \tag{8.3}$$

Again, it merely expresses the fact that θ^* is the expected value of z. This is called an *orthogonality* of *moment condition*. Let us now move to samples. Define

$$g_T(\theta) = E_T[z] - \theta \tag{8.4}$$

The sample counterpart of equation (8.3) is to find $\hat{\theta}$ that solves

$$g_T(\theta) = 0 \tag{8.5}$$

This obviously returns our estimator $\hat{\theta}$ from (8.2), namely, the sample moment or sample average.

8.2 ▪ Example of an Overidentified Model

With parametric models, one can run easily into overidentified models. This example is borrowed from Hamilton [19]. Suppose that $z(\omega)$ has the t density with θ degrees of freedom, namely,

$$f(z) = \frac{\Gamma[(\theta+1)/2]}{(\pi\theta)^{1/2}\Gamma(\theta/2)}(1 + z^2/\theta)^{-(\theta+1)/2}$$

It can then be proved that

$$E[z^2] = \frac{\theta}{\theta - 2} \tag{8.6}$$

$$E[z^4] = \frac{3\theta^2}{(\theta-2)(\theta-4)} \tag{8.7}$$

There are now two *moment conditions*: (8.6) and (8.7). Solving the first moment condition, (8.6), results in the following estimator $\hat{\theta}$:

$$\hat{\theta} = \frac{2E_T[z^2]}{E_T[z^2] - 1} \tag{8.8}$$

Solving the second moment condition (8.7) results in

$$\hat{\theta} = \frac{6E_T[z^4] \pm \sqrt{36E_T[z^4]^2 - 32E_T[z^4](E_T[z^4] - 3)}}{2(E_T[z^4] - 3)} \tag{8.9}$$

How do we mix the two? Clearly, both (8.8) and (8.9) cannot hold simultaneously. Indeed, we have $K = 1$ parameter and $J = 2$ moment conditions. Our model is overidentified.

Definition 8.1. *Let K be number of parameters to be estimated, and J the number of moment conditions. A system is* ***overidentified*** *if $J > K$. It is* ***just-identified*** *(or exactly identified) if $J = K$, and* ***underidentified*** *if $J < K$.*

The obvious solution is to replace the root-finding problem by the problem of minimizing a convex function J of θ. Let us define a 2-by-2 positive definite weight matrix $\mathbf{W}$. The sample errors are

$$\mathbf{g}_T(\theta) = \begin{bmatrix} E_T[z^2] - \frac{\theta}{\theta-2} \\ E[z^4] - \frac{3\theta^2}{(\theta-2)(\theta-4)} \end{bmatrix} \tag{8.10}$$

The GMM problem is to minimize over θ the weighted sum of the sample errors:

$$J_T(\theta) \equiv \frac{1}{2}\mathbf{g}_T(\theta)'\mathbf{W}\mathbf{g}_T(\theta) \tag{8.11}$$

The solution is given by setting the gradient equal to zero, i.e.,

$$\frac{d\mathbf{g}_T(\theta)'}{d\theta}\mathbf{W}\mathbf{g}_T(\theta) = 0 \tag{8.12}$$

Observations.

(1) In this example, the weighting matrix $\mathbf{W}$ is a "free parameter." We shall see how to exploit this fact in Chapter 25.

(2) In GMM, the errors $\mathbf{g}_T(\theta)$ are not necessarily linear functions of the parameters as (8.10) shows.

8.3 ▪ OLS as a Special Case of GMM

The two examples above were examples of unsupervised learning. We now look at the problem of linear regression, a classical supervised learning problem. The data $\mathbf{z}$ is split into the dependent variable y and the independent variable $\mathbf{x}$:

$$\mathbf{z} = \begin{bmatrix} y \\ \mathbf{x} \end{bmatrix}$$

The errors are linear functions of the parameters. OLS is thus a special case of GMM. What remains to be done is essentially a mapping from the notation of OLS into the notation of GMM; see Table 8.1.

Notation	OLS	GMM
Parameter	$\boldsymbol{b}$	$\boldsymbol{\theta}$
Optimal parameter	$\boldsymbol{\beta}$	$\boldsymbol{\theta}^*$

Table 8.1: Notation for OLS and GMM.

We recall that the population error in OLS is denoted by u:

$$u(\omega) = y(\omega) - \mathbf{x}'(\omega)\boldsymbol{\beta}$$

Half the gradient of the squared error is

$$\mathbf{p}(\mathbf{z}, \boldsymbol{\theta}) = \frac{1}{2}\frac{d(u^2)}{d\boldsymbol{\theta}}$$

Elementary calculus shows that

$$\mathbf{p}(\mathbf{z}(\omega), \boldsymbol{\theta}) = u(\omega)\mathbf{x}(\omega)$$

With that definition, the optimal parameter satisfies the orthogonality conditions $E[\mathbf{p}(\mathbf{z}, \theta^*)] = 0$. It is very tempting to define g_T as the average sample error $g_T = E_T[u]$, like in Section 8.1. Exercise 8.2 will show you why this choice (although perfectly justifiable statistically) does not result in an OLS formulation. The definition of g_T that corresponds to OLS is the sample scalar product:

$$\mathbf{g}_T(\theta) = E_T[u\mathbf{x}] \tag{8.13}$$

The goal of GMM is now to solve the (sample) orthogonality condition, which is a system of K equations with K unknowns:

$$\mathbf{g}_T(\theta) = \mathbf{0} \tag{8.14}$$

which is nothing but the normal equations (6.5), as you will see in Exercise 8.3.

Example 8.2. Returning to the Fama–French model (Example 6.4), we had $K = 4$ explanatory variables: a constant intercept $x_1 = 1$, the market return x_2, the return of small minus big stocks (SMB) x_3, and the return of high book/market minus low (HML) book/market stocks x_4. We have then four moment conditions. As the number of parameters is $K = 4$, the system can be identified perfectly. Trying to give names to these sample moments,

$$\mathbf{g}_T(\boldsymbol{\theta}) = \begin{bmatrix} E_T[u] \\ E_T[ux_2] \\ E_T[ux_3] \\ E_T[ux_4] \end{bmatrix} = \begin{bmatrix} \text{intercept condition} \\ \text{market return condition} \\ \text{SMB condition} \\ \text{HML condition} \end{bmatrix}$$

8.4 ▪ Vector OLS as a Special Case of GMM

What if we had a vector of dependent variables? That is, instead of regressing one stock on K factors, we are interested in regressing two stocks, with returns $y_{t,1}$ and $y_{t,2}$ on K factors $\mathbf{x}_t$. Defining the regression error on the first stock,

$$\begin{aligned} \hat{u}_{t,1} &= y_{t,1} - \mathbf{x}_t'\mathbf{b}^1 \\ \hat{u}_{t,2} &= y_{t,2} - \mathbf{x}_t'\mathbf{b}^2 \end{aligned}$$

We define

$$\boldsymbol{\theta} = \begin{bmatrix} \mathbf{b}^1 \\ \mathbf{b}^2 \end{bmatrix} \qquad \mathbf{g}_T(\boldsymbol{\theta}) = \begin{bmatrix} \frac{1}{T}\sum_{t=1}^T \mathbf{x}_t\hat{u}_{t,1} \\ \frac{1}{T}\sum_{t=1}^T \mathbf{x}_t\hat{u}_{t,2} \end{bmatrix}$$

A natural formulation of the vector OLS problem is to

$$\min \sum_t (\hat{u}_{t,1})^2 + \sum_t (\hat{u}_{t,2})^2$$

Its solution is again by the moment conditions:

$$\mathbf{g}_T(\boldsymbol{\theta}) = \mathbf{0}$$

8.5 ▪ Nonlinear Extensions

As we saw, GMM seeks a solution $\hat{\boldsymbol{\theta}}$ of the set of equations

$$\mathbf{A}_T \mathbf{g}_T(\boldsymbol{\theta}) = \mathbf{0} \tag{8.15}$$

For instance, in Sections 8.1, 8.3, and 8.4, we chose for $\mathbf{A}_T$ the identity matrix, while in Section 8.2, it was given by $\mathbf{A}_T = \mathbf{D}_T \mathbf{W}$, where $\mathbf{D}_T = \frac{d\mathbf{g}_T(\boldsymbol{\theta})}{d\boldsymbol{\theta}}$.

In the special case where $\mathbf{A}_T$ does not depend on the parameter $\boldsymbol{\theta}$, this method coincides (at least in empirical asset pricing) with a reformulation of OLS or GLS. Otherwise, like in Section 8.2, the problem can in general not be solved analytically, and one must resort to numerical techniques, which we turn to in the next chapters.

One of the reasons for the popularity of GMM among econometricians is that the distribution theory of $\hat{\boldsymbol{\theta}}$ and $\mathbf{g}_T(\hat{\boldsymbol{\theta}})$ has been elaborated upon by Hansen [20]. Indeed, it can be proved that $\hat{\boldsymbol{\theta}}(\omega)$ is unbiased, and its covariance is given by

$$TCov[\hat{\boldsymbol{\theta}}] = (\mathbf{AD})^{-1}\mathbf{ASA}'(\mathbf{AD})^{-1'}$$

where $\mathbf{S} = \sum_j \mathbf{p}'(\mathbf{z}(t), \boldsymbol{\theta}^*)\mathbf{p}(\mathbf{z}(t-j), \boldsymbol{\theta}^*)$, and $\mathbf{A}$ and $\mathbf{D}$ are the population equivalent of $\mathbf{A}_T$ and $\mathbf{D}_T$. For more information on GMM, we refer the reader to Hamilton [19], Hansen [20], and Cochrane [21].

Exercises

8.1. (Medium) Suppose your data is normally distributed with mean μ and variance σ^2. Construct a GMM estimator of μ using the first three sample moments.

8.2. (Hard) For the OLS problem in Section 8.3, take $\mathbf{g}_T(\boldsymbol{\theta}) = E_T[\mathbf{u}]$, i.e., the mean sample error. Show that solving the problem

$$\min \mathbf{g}_T(\boldsymbol{\theta})'\mathbf{g}_T(\boldsymbol{\theta})$$

does not result in the estimator given by the normal equations.

8.3. (Medium) Refer to the definition (8.13) of $\mathbf{g}_T(\theta)$ for the OLS problem in Section 8.3. Show why it does not hurt in OLS to present GMM as a minimizing method. In other words, show that

$$\min \mathbf{g}_T(\boldsymbol{\theta})'\mathbf{g}_T(\boldsymbol{\theta})$$

results in the estimator given by the normal equations.

8.4. (Easy) Cast GLS in the GMM framework, i.e., define $\mathbf{A}_T$ and $\mathbf{g}_T(\boldsymbol{\theta})$ such that GLS solves (8.15).

Chapter 9

Nonlinear Optimization Algorithms

The convention in most books on optimization is to consider **minimization** problems instead of maximization problems, and we will adhere to it in this chapter.

Unlike quadratic optimization algorithms, of which the mean-variance problem is an egregious example, most optimization problems cannot be solved analytically. In that case, one must resort to numerical methods or algorithms. Most algorithms are iterative, i.e., they start at an initial value $\mathbf{x}^0$ and recursively calculate a new iterate $\mathbf{x}^{k+1}$ based on the information at time k, i.e., $\mathbf{x}^k$ and $f(\mathbf{x}^k)$, and possibly its derivatives, until a termination criterion is met. Optimization algorithms can be classified into

- unconstrained optimization
- constrained optimization.

In unconstrained optimization, the goal is to minimize a function $f(\mathbf{x})$, whereas in constrained optimization, the additional goal is to satisfy constraints of the form $\mathbf{g}(\mathbf{x}) = \mathbf{0}$.

There are several very clever methods to solve constrained optimization problems. In neural network learning, the introduction of the sigmoid function results in bounding some variables automatically, and thus constraints are often no longer necessary. More generally, a **hard constraint** $\mathbf{g}(\mathbf{x}) = \mathbf{0}$ can be transformed into a soft constraint; i.e., the objective f can be replaced by a new objective L with

$$L(\mathbf{x}) = f(\mathbf{x}) + \lambda||\mathbf{g}(\mathbf{x})||$$

With the right choice of $\lambda > 0$, the constraint $\mathbf{g}(\mathbf{x}) = 0$ is set close to zero and $f(\mathbf{x})$ is minimized. This technique is called **regularization**. From now on, we will focus on unconstrained optimization, in particular, the minimization problem

$$\min_{\mathbf{x}} f(\mathbf{x}) \tag{9.1}$$

When the problem is nonconvex,[24] no algorithm except brute search is guaranteed to converge to a global minimum. However, most algorithms will converge to a stationary point, i.e., a point where

$$||\nabla f(\mathbf{x}^k)|| \leq \varepsilon \tag{9.2}$$

Equation (9.2) is often used as a termination criterion. Thus, for optimization problems where we can calculate exactly the derivatives of f at every step (such as neural networks), a general

[24]A *convex* problem is a problem where f is quasiconvex, i.e., when, over a segment, the maximum of the function is attained at the endpoint(s); quasiconvex functions generalize convex functions.

minimization algorithm is the descent algorithm where, at every step, one makes sure that the function decreases.[25]

ALGORITHM 9.1. Descent with derivative algorithm.

1. Set $k = 0$, $\mathbf{x}^0$ =any value
2. Repeat
   Choose a vector $\mathbf{x}^{k+1}$ such that $f(\mathbf{x}^{k+1}) < f(\mathbf{x}^k)$
   $k \leftarrow k+1$
3. Until $||\nabla f(\mathbf{x}^k)|| \leq \varepsilon$

Other termination criteria can be used, such as limiting the number of iterations.

We now distinguish between two types of methods:

- first-order optimization algorithms
- second-order optimization algorithms.

As the name indicates, first-order (second-order) optimization algorithms consider the first-order (second-order) approximation of the function f.

9.1 ▪ First-Order Descent Algorithms

9.1.1 ▪ The Scalar Case

Let us first consider the scalar case. By Taylor approximation

$$f(x^{k+1}) \approx f(x^k) + \frac{df}{dx}(x^k)(x^{k+1} - x^k)$$

The only useful part of this equation is that it tells us (based on the sign of the derivative $\frac{df}{dx}(x^k)$) whether x^{k+1} should be larger or smaller than x^k to satisfy our descent condition. As Figure 9.1 shows, one can overshoot the minimum.

9.1.2 ▪ The Multivariate Case

In several dimensions, the latter observation becomes more interesting. The Taylor approximation is indeed

$$f(\mathbf{x}^{k+1}) \approx f(\mathbf{x}^k) + \nabla f(\mathbf{x}^k)^t(\mathbf{x}^{k+1} - \mathbf{x}^k)$$

Thus the direction opposite to the gradient, i.e.,

$$\mathbf{x}^{k+1} - \mathbf{x}^k = -\eta \nabla f(\mathbf{x}^k) \tag{9.3}$$

for small $\eta \geq 0$ is guaranteed to give us the largest decrease of the linear approximation of the function f if the latter is twice differentiable. Indeed, let us take any another candidate direction $\mathbf{p}$, with the same length as the gradient, i.e.,

$$||\mathbf{p}|| = ||\nabla f(\mathbf{x}^k)||$$

[25] One must be sure also that the function decreases "sufficiently"; this condition is beyond the scope of this book.

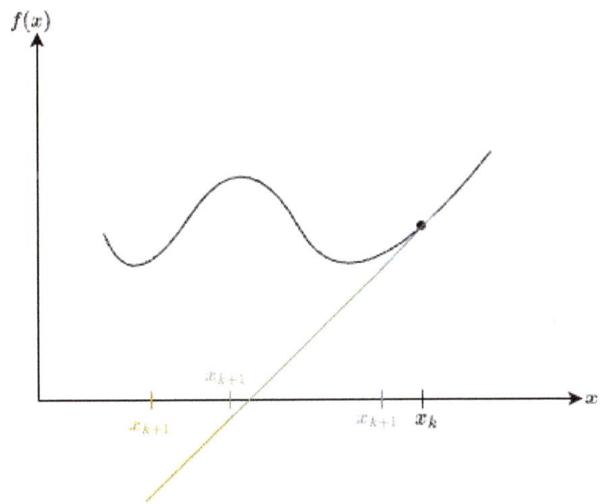

Figure 9.1: Unless one uses second-order information, it is unclear whether the blue/green or orange x_{k+1} is the correct next step.

Then, by the Cauchy–Schwarz inequality,

$$-\nabla f(\mathbf{x}^k)^t \cdot \nabla f(\mathbf{x}^k) \leq -\nabla f(\mathbf{x}^k)^t \cdot \mathbf{p}$$

Thus, choosing the (opposite) direction to the gradient (i.e., (9.3)) results in a larger decrease in the objective function than choosing any other direction, $\mathbf{p}$, as Figure 9.2 shows.

Unfortunately, it is difficult to know what should be the largest value of the **learning rate**, or *step length*, η such that gradient descent does not "overshoot" the objective (see Figure 9.1 again). Sometimes, a line search algorithm is used to determine η. The idea is to solve the

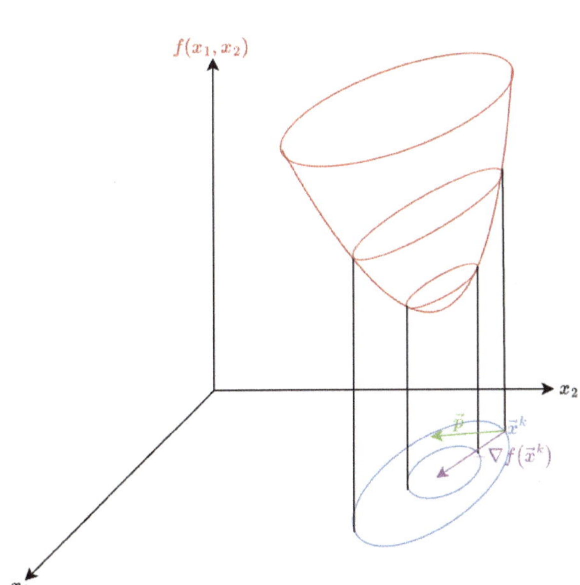

Figure 9.2: (Minus) the gradient points to a better direction than any other direction $\mathbf{p}$.

single-variable problem

$$\min_{\eta} f(\mathbf{x}^k - \eta \nabla f(\mathbf{x}^k)) \tag{9.4}$$

While the line search problem seems identical to problem (9.1), it is fundamentally different and much simpler computationally, as it is a univariate optimization problem. It can thus be solved by exhaustive search.

To summarize, the following gradient descent algorithm (also called steepest descent) is guaranteed to converge to a strict minimum of f if f is convex and twice differentiable, as ϵ goes to zero.

ALGORITHM 9.2. Gradient descent with line search algorithm.

1. Set $k=0$, $\mathbf{x}^0$ =any value
2. Repeat
   $\eta_k = \arg\min_\eta f(\mathbf{x}^k - \eta\nabla f(\mathbf{x}^k))$
   Set $\mathbf{x}^{k+1} = \mathbf{x}^k - \eta_k \nabla f(\mathbf{x}^k)$
   $k \leftarrow k+1$
3. Until $||\nabla f(x^k)|| \leq \varepsilon$

9.2 ▪ Second-Order Descent Algorithms

9.2.1 ▪ The Scalar Case

Let us first analyze Newton's method in the scalar case. By second-order Taylor approximation,

$$f(x^{k+1}) \approx f(x^k) + \frac{df}{dx}(x^k)(x^{k+1} - x^k) + \frac{1}{2}\frac{d^2 f}{dx^2}(x^k)(x^{k+1} - x^k)^2$$

In order to minimize $f(x^{k+1})$ is is easy to see that one must minimize the "quadratic approximation q of f at x^{k+1}," i.e.,

$$\min_{x_{k+1}} q(x_{k+1}) \equiv \frac{df}{dx}(x^k)(x^{k+1} - x^k) + \frac{1}{2}\frac{d^2 f}{dx^2}(x^k)(x^{k+1} - x^k)^2 \tag{9.5}$$

The first-order condition $\frac{dq}{dx}|_{x^{k+1}} = 0$ is

$$\frac{df}{dx}(x^k) + \frac{d^2 f}{dx^2}(x^k)(x^{k+1} - x^k) = 0 \tag{9.6}$$

Thus

$$x^{k+1} = x^k - \frac{\frac{df}{dx}(x^k)}{\frac{d^2 f}{dx^2}(x^k)}$$

This shows that the learning rate should be the inverse of the curvature:

$$\eta = \frac{1}{\frac{d^2 f}{dx^2}(x^k)}$$

An obvious problem seems to plague the method: what if $\frac{d^2 f}{dx^2}(x^k) < 0$? Looking at (9.5), it is clear that the stationary point of q will be a maximum; thus the Newton step is likely to *increase* f, not decrease it. This does not occur when f is convex.

9.2.2 ▪ The Multivariate Case

The second-order Taylor approximation is

$$f(\mathbf{x}^{k+1}) \approx f(\mathbf{x}^k) + \nabla f(\mathbf{x}^k)'(\mathbf{x}^{k+1} - \mathbf{x}^k) + \frac{1}{2}(\mathbf{x}^{k+1} - \mathbf{x}^k)'\frac{d^2 f}{d\mathbf{x}^2}(\mathbf{x}^k)(\mathbf{x}^{k+1} - \mathbf{x}^k) \quad (9.7)$$

In order to minimize $f(\mathbf{x}^{k+1})$ one must minimize the "quadratic approximation q of f at $\mathbf{x}^k$," i.e.,

$$\min_{x_{k+1}} q(\mathbf{x}^{k+1}) \equiv \nabla f(\mathbf{x}^k)'(\mathbf{x}^{k+1} - \mathbf{x}^k) + \frac{1}{2}(\mathbf{x}^{k+1} - \mathbf{x}^k)'\frac{\partial^2 f}{\partial \mathbf{x}^2}(\mathbf{x}^k)(\mathbf{x}^{k+1} - \mathbf{x}^k) \quad (9.8)$$

The first-order condition $\frac{dq}{d\mathbf{x}}|_{\mathbf{x}^{k+1}} = 0$ is

$$\nabla f(\mathbf{x}^{k+1}) + \frac{\partial^2 f}{\partial \mathbf{x}^2}(\mathbf{x}^k)(\mathbf{x}^{k+1} - \mathbf{x}^k) = 0 \quad (9.9)$$

Thus

$$\mathbf{x}^{k+1} = \mathbf{x}^k - \left(\frac{\partial^2 f}{\partial \mathbf{x}^2}(\mathbf{x}^k)\right)^{-1} \nabla f(\mathbf{x}^{k+1})$$

We then obtain Newton's algorithm.

ALGORITHM 9.3. Newton's algorithm.

1. Set $k = 0$, $\mathbf{x}^0$ =any value
2. Repeat
   Set $\mathbf{x}^{k+1} = \mathbf{x}^k - \left(\frac{\partial^2 f}{\partial \mathbf{x}^2}(\mathbf{x}^k)\right)^{-1} \nabla f(\mathbf{x}^k)$
   $k \leftarrow k + 1$
3. Until $||\nabla f(\mathbf{x}^k)|| \leq \varepsilon$

The same (potential) problem plagues Newton's method in the vector case as in the scalar case. When the Hessian $\frac{\partial^2 f}{\partial \mathbf{x}^2}(\mathbf{x}^k)$ is not strictly positive definite, then anything can happen. Note that many algorithms, such as the Davidon–Fletcher–Powell (DFP) and the Broyden–Fletcher–Goldfarb–Shanno (BFGS) algorithms, have been invented to circumvent that problem in low dimensions. It is easy to see (see Exercise 9.6) that the proportion of saddle points (over the population of stationary points) increases with the size of the problem. Thus ML algorithms, meant for big data, will not use Newton's algorithm since it is likely to be stuck in a saddle point. A potential remedy to that problem is simulated annealing. Another problem of Newton's algorithm is that the Hessian needs to be inverted. When the system is large, this is computationally costly and potentially error-prone, like the inversion of the variance-covariance matrix in mean-variance. (Look at how closely the problems are related!)

Finally, let us use Hessian information to determine an adaptive learning rate η in steepest descent. Substituting (9.3) in (9.8), we obtain

$$q(\eta) = -\eta \nabla f(\mathbf{x}^k)'\nabla f(\mathbf{x}^k) + \frac{1}{2}\eta^2 \nabla f(\mathbf{x}^k)'\frac{\partial^2 f}{\partial \mathbf{x}^2}(\mathbf{x}^k)\nabla f(\mathbf{x}^k)$$

Minimizing q with respect to η, we obtain an adaptive learning rate:

$$\eta^k = \frac{\nabla f(\mathbf{x}^k)'\nabla f(\mathbf{x}^k)}{\nabla f(\mathbf{x}^k)'\frac{\partial^2 f}{\partial \mathbf{x}^2}(\mathbf{x}^k)\nabla f(\mathbf{x}^k)} \quad (9.10)$$

Again, the advantage of this type of gradient descent over Newton's method is that we do not need to invert the Hessian, even if we use formula (9.10).

9.3 ▪ Nonlinear Least Squares Problems

As we have seen, least squares problems are particularly important in machine learning and asset pricing. Indeed, nonlinear least squares are the problem that supervised learning tries to solve. It is instructive to try to tailor our algorithms to nonlinear least squares problems. We now return to ML notation in Table 9.1.

	Machine learning	Optimization
Objective	J	f
Variable	$\boldsymbol{\theta} \in \mathbb{R}^K$	$\mathbf{x}$

Table 9.1: Notation in ML and optimization.

In supervised learning, one tries to find a function $f^*(\mathbf{x})$ such that the intrinsic error $u \equiv y - f^*(\mathbf{x})$ is minimized. Due to approximation error, one can only find an approximation of the true function $f^*(\mathbf{x})$, which we denote by $f(\mathbf{x};\boldsymbol{\theta}^*)$. If we had an infinite amount of data, we might be able to find the optimal parameter $\boldsymbol{\theta}^*$. However, with a finite number of data (or an imprecise algorithm), the output of the algorithm is only an imprecise estimate $\hat{\boldsymbol{\theta}}$ of $\boldsymbol{\theta}^*$. We can thus decompose the error into

$$\hat{u}_t = y_t - f(\mathbf{x}_t;\hat{\boldsymbol{\theta}}) = \underbrace{f^*(\mathbf{x}_t) - f(\mathbf{x};\boldsymbol{\theta}^*)}_{\text{approximation error}} + \underbrace{f(\mathbf{x}_t;\boldsymbol{\theta}^*) - f(\mathbf{x}_t;\hat{\boldsymbol{\theta}})}_{\text{estimation error}} + \underbrace{u_t}_{\text{intrinsic error}}$$

The intrinsic error is genuinely unpredictable. The approximation error comes from the misspecification of the model and can be reduced by a more general specification. As we saw, however, this general specification may lead to generalization error, i.e., overfitting. The estimation error can be reduced by adding more observations (i.e., increasing T) or using a better algorithm.

In OLS, the model is

$$f(\mathbf{x}_t;\boldsymbol{\theta}) = \boldsymbol{\theta}'\mathbf{x}_t \tag{9.11}$$

so we assume the same number of features as parameters ($I = K$). More generally, the nonlinear least squares problem can be written as

$$\min_{\boldsymbol{\theta}} J(\boldsymbol{\theta}) \equiv \frac{1}{2}\sum_{t=1}^{T} \hat{u}_t(\boldsymbol{\theta})^2$$

Clearly,

$$\nabla J(\boldsymbol{\theta}) = \boldsymbol{X}(\boldsymbol{\theta})'\hat{\mathbf{u}}(\boldsymbol{\theta})$$

where

$$\mathbf{X}(\boldsymbol{\theta})' = -\begin{bmatrix} \frac{\partial f_1(\theta)}{\partial \theta_1} & \cdots & \frac{\partial f_T(\theta)}{\partial \theta_1} \\ \vdots & & \vdots \\ \frac{\partial f_1(\theta)}{\partial \theta_K} & \cdots & \frac{\partial f_T(\theta)}{\partial \theta_K} \end{bmatrix}$$

$$\hat{\mathbf{u}}(\boldsymbol{\theta}) = \begin{bmatrix} \hat{u}_1(\boldsymbol{\theta}) \\ \vdots \\ \hat{u}_T(\boldsymbol{\theta}) \end{bmatrix}$$

The Hessian of J is

$$\frac{d^2 J}{d\boldsymbol{\theta}^2}(\boldsymbol{\theta}) = \mathbf{X}(\boldsymbol{\theta})'\mathbf{X}(\boldsymbol{\theta}) + \mathbf{Q}(\boldsymbol{\theta})$$

where

$$\mathbf{Q}(\boldsymbol{\theta}) = \sum_{t=1}^{T} \hat{u}_t(\boldsymbol{\theta}) \frac{\partial^2 f(\mathbf{x}_t; \boldsymbol{\theta})}{\partial \boldsymbol{\theta}^2}$$

The $\mathbf{X}(\boldsymbol{\theta})'\mathbf{X}(\boldsymbol{\theta})$ term will typically dominate the term $\mathbf{Q}(\boldsymbol{\theta})$ (which will be zero for linear least squares). Interestingly, Belkin et al. [11] show that a feedforward neural network has $\mathbf{Q}(\boldsymbol{\theta}) \to \mathbf{0}$ when the number of parameters tends to infinity. However, the same is not true for the term $\mathbf{X}(\boldsymbol{\theta})'\mathbf{X}(\boldsymbol{\theta})$. This is similar to the diversification effect in mean-variance analysis, as we shall see in Chapter 16: $\mathbf{X}(\boldsymbol{\theta})$ resembles a mean effect, while $\mathbf{Q}(\boldsymbol{\theta})$ resembles a variance or risk effect. With an infinite amount of parameters correctly diversified, all "risk" is washed away.

The Newton direction is given by

$$\boldsymbol{\theta}^{k+1} - \boldsymbol{\theta}^k = -\left(\mathbf{X}(\boldsymbol{\theta}^k)'\mathbf{X}(\boldsymbol{\theta}^k) + \mathbf{Q}(\boldsymbol{\theta}^k)\right)^{-1} \mathbf{X}(\boldsymbol{\theta}^k)'\,\hat{\mathbf{u}}(\boldsymbol{\theta}^k)$$

In the Gauss–Newton algorithm,[26] one sets $\mathbf{Q} = \mathbf{0}$ and obtains the following.

ALGORITHM 9.4. Gauss–Newton Algorithm.

1. Set $k = 0$, $\boldsymbol{\theta}^0$ =any value
2. Repeat
   Set $\boldsymbol{\theta}^{k+1} = \boldsymbol{\theta}^k - \left(\mathbf{X}(\boldsymbol{\theta}^k)'\mathbf{X}(\boldsymbol{\theta}^k)\right)^{-1} \mathbf{X}(\boldsymbol{\theta}^k)'\,\hat{\mathbf{u}}(\boldsymbol{\theta}^k)$
   $k \leftarrow k + 1$
3. Until $||\nabla J(\mathbf{x}^k)|| \leq \varepsilon$

This algorithm is not used much in practice, but an extension of it, the Levenberg–Marquardt algorithm, is used a lot, as you will see in Exercise 9.3. Again, observe that when the function f is linear in $\boldsymbol{\theta}$, this algorithm will converge to the solution of the normal equations.

9.4 ▪ To Probe Further

This material is classical. We refer the reader to the books [15] and [16].

Exercises

9.1. Find the minimizer of the following functions by the gradient algorithm with $\eta = 1$ and Newton's algorithm. Start at $x^0 = 1$.

(1) $f(x) = x^3$

(2) $f(x) = x^4$

(3) $f(x) = (x-1)^2(x-2)^2$

What do you observe?

[26] An indirect consequence of [11] is that Gauss and Newton "invented" feedforward neural networks.

9.2. Find the minimizer of the following functions by the gradient algorithm with line search and Newton's algorithm. Start at

$$\mathbf{x}^0 = \begin{bmatrix} 1 \\ 1 \end{bmatrix}$$

(1) $f(x) = (x_1 + x_2)^3$

(2) $f(x) = x_1^4 + x_2^4$

(3) $f(x) = (x_1 + x_2 - 2)^2(x_1 - 4)^2$

Feel free to use brute search for the line search, i.e., vary the learning rate by very small steps and keep the value η_k^* that gives you the smallest value of $f(\mathbf{x}^k - \eta_k \nabla f(\mathbf{x}^k))$.

What do you observe?

9.3. In some problems, the dependent variable is positive. Use the Levenberg–Marquardt algorithm (with constant λ^k) to solve the following problem:

$$\min_{\theta_1,\theta_2} \sum_{t=1}^{2} (\exp(\theta_1 x_{t,1} + \theta_2 x_{t,2}) - y_t)^2$$

Verify your results by solving the following problem using the normal equations:

$$\min_{\theta_1,\theta_2} \sum_{t=1}^{2} (\theta_1 x_{t,1} + \theta_2 x_{t,2} - \ln y_t)^2$$

The algorithm and dataset are provided below.

ALGORITHM 9.5. Levenberg–Marquardt algorithm.

1. Set $k = 0$, $\boldsymbol{\theta}^0$ =any value, λ^0 =any small value
2. Repeat
   Set $\boldsymbol{\theta}^{k+1} = \boldsymbol{\theta}^k - (\mathbf{X}(\boldsymbol{\theta}^k)'\mathbf{X}(\boldsymbol{\theta}^k) + \lambda^k \mathbf{I})^{-1}\mathbf{X}(\boldsymbol{\theta}^k)'\hat{\mathbf{u}}(\boldsymbol{\theta}^k)$
   $k \leftarrow k + 1$
   Update λ^k
3. Until $||\nabla J(\boldsymbol{\theta}^k)|| \le \varepsilon$

Note that you must figure out for yourself what J is. Use for data:

t	$x_{t,1}$	$x_{t,2}$	y_t
1	0.5	0.6	3
2	0.8	0.9	4

9.4. (Hard) Consider the ordinary linear least squares problem, where the design matrix $\mathbf{X}$ is square, and

$$\mathbf{X} = \mathbf{I} - \frac{\delta}{2}\mathbf{v}^t\mathbf{v}$$

for some unit length vector $\mathbf{v}$, and δ a small number.

(a) Prove the Neumann approximation

$$(\mathbf{X}'\mathbf{X})^{-1} = \mathbf{I}+\delta\mathbf{v}^t\mathbf{v}+O(\delta^2)$$

(b) For what value of the learning rate will the gradient descent algorithm (without line search) converge?

9.5. Index tracking consists in replicating an index with minimum error. This time we will replicate the S&P 500 index by minimizing error

$$\min_{\mathbf{b}} \sum (r_{M,t} - b_1 r_{G,t} - b_2 r_{W,t} - b_3 r_{A,t} - b_4 r_{GM,t} - b_5 r_{I,t} - b_6 r_{P,t})^2$$

where $r_{i,t}$ represents the returns for the following table:

Symbol	Company
$r_{M,t}$	S&P 500
$r_{G,t}$	Google
$r_{W,t}$	Walmart
$r_{A,t}$	Apple
$r_{GM,t}$	General Motors (GM)
$r_{I,t}$	Intel
$r_{P,t}$	Pfizer

(a) Implement in either MATLAB, R, or Python the gradient descent algorithm (Algorithm 9.2). You need to write the code yourself using the notation in this book instead of calling a routine from a library. For the line search, feel free to choose brute search, i.e., vary the learning rate by very small steps and keep the value η_k^* that gives you the smallest value of $f(\mathbf{x}^k - \eta_k \nabla f(\mathbf{x}^k))$.

(b) Use as a training period from April 1st to August 1st, 2021. Assess the performance of your method on the test period from August 1st, 2021, to October 1st, 2021. The data is provided in the supplementary material.[27]

9.6. (Medium) Explain why the proportion of saddle points of a function (out of all stationary points) generically increases with the dimension of the argument.

[27]https://bookstore.siam.org/ot195/bonus

Chapter 10

Unsupervised Learning

Unsupervised learning has its roots in the classical problem of multivariate statistics: how to estimate a joint probability distribution $P(\tilde{\mathbf{x}} \leq \mathbf{x})$, where $\tilde{\mathbf{x}}$ denotes a vector-valued random variable in I dimensions. We will give a quick (and thus necessarily) incomplete summary of classical density estimation before turning to methods for higher dimensions, principal component analysis, and clustering.

Before turning to these methods, it is worth emphasizing that the line between unsupervised and supervised learning is often blurred. There are correspondences.

10.1 ▪ Correspondence with Supervised Learning

The idea is as follows. Suppose that the number of dimensions of each data point is $I = 2$. We write the (realization of the) random vector:

$$\tilde{\mathbf{x}}(\omega) = \begin{bmatrix} \tilde{x}_1(\omega) \\ \tilde{x}_2(\omega) \end{bmatrix}$$

We arbitrarily attribute two different roles to each of the scalar variables $\tilde{x}_1(\omega)$ and $\tilde{x}_2(\omega)$, assumed dependent (in the probability sense). The variable $\tilde{x}_1(\omega)$ will be the independent variable, in the sense of supervised learning, while the variable $\tilde{x}_2(\omega)$ will be the dependent variable. Calling $(x_{1,1}, \ldots, x_{1,T})$ the observations of $\tilde{x}_1(\omega)$ and $(x_{2,1}, \ldots, x_{2,T})$ the observations of $\tilde{x}_2$, we then solve the minimization problem over our training dataset:

$$\min_{\boldsymbol{\theta}} \frac{1}{T} \sum_{t=1}^{T} (x_{2,t} - f(x_{1,t}, \boldsymbol{\theta}))^2 \tag{10.1}$$

which is nothing but the classical (scalar) nonlinear regression problem, as in Chapter 9, where the independent observations were written x_t (instead of $x_{1,t}$) and the dependent observations were written y_t (instead of $x_{2,t}$). The following theorem states that the solution of the (population) regression problem is given by the conditional expectation.

Theorem 10.1 (nonlinear estimation theorem). *The function that minimizes* $E[(\tilde{x}_2 - f(\tilde{x}_1))^2]$ *is*

$$f^*(\tilde{x}_1(\omega)) = E[\tilde{x}_2|\tilde{x}_1](\omega) \tag{10.2}$$

Equation (10.2) is the fundamental principle of **nonlinear estimation theory**.[28] The mathematically inclined reader is encouraged to do Exercise 10.1, where he/she will be guided through the steps of the proof of that principle.

You will see in Exercise 10.5 a theoretical method, based on this principle, to infer the empirical conditional probability distribution

$$P(\tilde{x}_2 \le x_2 | \tilde{x}_1 = x_1) \tag{10.3}$$

This method is not efficient. For that reason, statisticians prefer to work with distributions that are characterized by few parameters, like the Gaussian distribution, and to estimate the moments. While the conditional distribution (10.3) is not sufficient to determine the joint distribution of $\tilde{\mathbf{x}}$, this (supervised) approach is often useful for density estimation of **stochastic processes**, as the following example shows.

Example 10.2. Let us consider a scalar stochastic process $\{\tilde{x}_i\}$ with the Markov property.[29] Suppose the process takes values in the positive integers. Then the following equality holds:

$$P(\tilde{x}_I = x_I, \ldots, \tilde{x}_1 = x_1 | \tilde{x}_0 = x_0) \tag{10.4}$$

$$= \sum_{x_{I-1}=1}^{\infty} \cdots \sum_{x_1=1}^{\infty} P(\tilde{x}_I = x_I | \tilde{x}_{I-1} = x_{I-1}) \ldots P(\tilde{x}_1 = x_1 | \tilde{x}_0 = x_0) \tag{10.5}$$

Recall that in finance there is only one sample point of the process. Thus, for estimation, one usually assumes *stationarity* of the conditional probability mass functions:

$$P(\tilde{x}_1 = y | \tilde{x}_0 = x) = P(\tilde{x}_{i+1} = y | \tilde{x}_i = x) \qquad i = 1, \ldots, I-1$$

Usually, the value of the process at time zero, namely, x_0, is known, so (10.4) is the joint distribution.

10.2 ▪ Density Estimation in Finance

In low-dimensional problems, one can use the same techniques as in the univariate case. For parametric models, *maximum likelihood estimation* (MLE) is popular among statisticians for its theoretical properties. MLE can be viewed as the methodology of choice for unsupervised learning, while the (generalized) method of moments is the methodology for regressions. They are known to be equivalent for some distributions (see Exercise 3.4), thus bolstering the view that the line between supervised and unsupervised learning is blurred.

We will see an application of MLE to a time series model with only $K = 3$ parameters in Chapter 12. However, MLE often does not scale well to high-dimensional financial problems [14]. For nonparametric models, classical estimation of the empirical distribution works well for small dimensions (say, less than 3). See Silverman [22].

For higher dimensions, the curse of dimensionality manifests itself: when the number of dimensions I or the number of parameters K in a parametric model is not significantly smaller

[28]The reader is referred to Figure 6.2 in Chapter 6 on OLS. It illustrates that, for linear functions f, the optimal estimator f^* is the projection of the dependent variables on the space spanned by the independent variables. This is exactly what the conditional expectation is. The same principle holds for nonlinear estimators.

[29]In the remainder of the book we use the index t for time; in this example we use the index i as we consider a realization of the process to be a vector in $\mathbb{R}^I$. The reader, if puzzled, is urged to read our discussion on ergodicity.

than the number T of observations $\{\mathbf{x}_1, \ldots, \mathbf{x}_T\}$, one can resort to two different approaches:

- enforce restrictions on the class of (parametric) models one wants to estimate
- decrease the dimension by nonparametric (data analytic) approaches.

As an example of the parametric approach, the **copula** methodology (see Sklar et al. [23]) was quite popular in financial risk management between 2000 and 2010.[30] In the copula approach, one starts by inferring marginal distributions, for which the number of observations T is generally sufficient to provide nonspurious estimators. Then one proceeds to merge these marginals into a joint distribution through a copula function. For instance, if all marginals were Gaussian, and one uses a Gaussian copula, one would obtain a Gaussian joint distribution. However, one can fit other marginals than the Gaussian.

For more information on classical density estimation, the reader is referred to Hastie, Tibsharini, and Friedman [24]. We now turn to two popular data analytic approaches: principal component analysis and clustering.

10.3 ▪ Principal Component Analysis (PCA)

We saw the theory of PCA in Exercise 3.9. Let us look at this method in more detail. In finance, returns are often IID and are often the (vector) variable of choice for analysis. For that reason, we use the symbol $\mathbf{r}$ (rather than $\mathbf{x}$) for our random vector:

$$\mathbf{r}(\omega) = \begin{bmatrix} r_1(\omega) \\ \vdots \\ r_i(\omega) \\ \vdots \\ r_I(\omega) \end{bmatrix}$$

The idea is to rotate our vector $\mathbf{r}(\omega)$ into a vector $\mathbf{x}(\omega)$ so that x_i and x_j are uncorrelated random variables. More specifically, the idea is to find a unitary matrix $\mathbf{W}$ such that

$$\mathbf{x}(\omega) = \mathbf{W}\mathbf{r}(\omega)$$

Let

$$\mathbf{W} = \begin{bmatrix} \mathbf{w}^1 \\ \vdots \\ \mathbf{w}^n \end{bmatrix}$$

This can be done by the following procedure:

- The first principal component of $\mathbf{r}$ is the linear combination $x_1(\omega) = \mathbf{w}^1\mathbf{r}(\omega)$ that maximizes $Var[x_1]$ subject to the constraint $||\mathbf{w}^1||_2 = 1$.
- The second principal component of $\mathbf{r}$ is the linear combination $x_2(\omega) = \mathbf{w}^2\mathbf{r}(\omega)$ that maximizes $Var[x_2]$ subject to the constraint $||\mathbf{w}^2||_2 = 1$ and $Cov[x_1, x_2] = 0$.
- The pth principal component of $\mathbf{r}$ is the linear combination $x_p(\omega) = \mathbf{w}^p\mathbf{r}(\omega)$ that maximizes $Var[x_p]$ subject to the constraint $||\mathbf{w}^p||_2 = 1$ and $Cov[x_p, x_j] = 0$ for $j = 1, \ldots, p-1$.

[30]The Gaussian copula approach was indeed so popular in industry that it was accused of misrepresenting the risk of subprime mortgages in a systemic way, and thus leading to the subprime crisis of 2008.

Let us now make the link with spectral decomposition. Suppose we know the variance-covariance matrix $\mathbf{\Sigma}$ of returns, i.e.,

$$\mathbf{\Sigma} = \begin{bmatrix} Var[r_1] & \dots & Cov[r_1, r_i] & \dots & Cov[r_1, r_I] \\ \vdots & & \vdots & & \vdots \\ Cov[r_1, r_i] & \dots & Var[r_i] & \dots & \vdots \\ \vdots & & \vdots & & \vdots \\ Cov[r_1, r_I] & \dots & \dots & \dots & Var[r_I] \end{bmatrix}$$

Since the variance-covariance matrix $\mathbf{\Sigma}$ is nonnegative definite, it has a spectral decomposition:

$$\mathbf{\Sigma} = \mathbf{W \Lambda W}'$$

with $\mathbf{W}$ (the eigenvector matrix) unitary and $\mathbf{\Lambda}$ (the eigenvalue matrix) diagonal with nonnegative entries $\{\lambda_p\}$. As you saw in Exercise 3.9, the eigenvector matrix is identical to the rotation matrix $\mathbf{W}$ in Section 10.3. Thus,

$$Var[x_p] = \lambda_p$$

Since the trace of $\mathbf{\Sigma}$ is the same as the trace of $\mathbf{W\Lambda W}'$, we have

$$\sum_{i=1}^{I} Var[r_i] = \sum_{i=1}^{I} Var[x_i]$$

This can be written as

$$\frac{Var[x_p]}{\sum_{i=1}^{I} Var[r_i]} = \frac{\lambda_p}{\lambda_1 + \cdots + \lambda_I}$$

Consequently, the proportion of the total variance in $\mathbf{r}$ explained by the pth principal component is the ratio between the pth eigenvalue and the trace of $\mathbf{\Sigma}$.

Example 10.3 (stocks). This example is borrowed from Tsay [25]. "Consider the monthly log returns from International Business Machines, Hewlett-Packard, Intel Corporation, Merrill Lynch, and Morgan Stanley Dean Witter from January 1990 to December 1999. The returns are in percentages and include dividends." The dataset has $T = 120$ observations. As expected, returns of companies in the same industrial sector tend to exhibit similar patterns. The sample mean is

$$\bar{\mathbf{r}} = \begin{bmatrix} \bar{r}_{IBM} \\ \bar{r}_{HPQ} \\ \bar{r}_{INTC} \\ \bar{r}_{MER} \\ \bar{r}_{MWD} \end{bmatrix} = \begin{bmatrix} 1.47 \\ 1.97 \\ 3.05 \\ 2.3 \\ 2.36 \end{bmatrix}$$

and the variance-covariance matrix is

$$\mathbf{\Sigma} = \begin{bmatrix} 73.10 & 36.48 & 27.09 & 16.06 & 16.33 \\ 36.48 & 103.6 & 48.86 & 37.59 & 40.72 \\ 27.09 & 48.86 & 113.96 & 27.96 & 6.86 \\ 16.06 & 37.59 & 27.96 & 105.56 & 85.47 \\ 16.33 & 40.72 & 26.86 & 85.47 & 109.91 \end{bmatrix}$$

The result of PCA is

Eigenvalue	256.16	116.14	64.91	46.82	22.11
Eigenvector	0.246	0.327	0.586	−0.7	0.018
	0.461	0.360	0.428	0.687	−0.05
	0.409	0.585	−0.683	−0.153	0.033
	0.522	−0.452	−0.082	−0.115	−0.710
	0.536	−0.467	−0.036	−0.042	0.701

The first component is a roughly equally weighted linear combination of the stock returns. This is called the *market component.* The second component represents the difference between the two industrial sectors—namely, technologies vs. financial services. It might be called an *industrial component.*

Example 10.4 (bonds). This example is borrowed from Fabozzi [26]. It describes the US Treasury yield curve. The dimension I of the data is the number of different maturities.

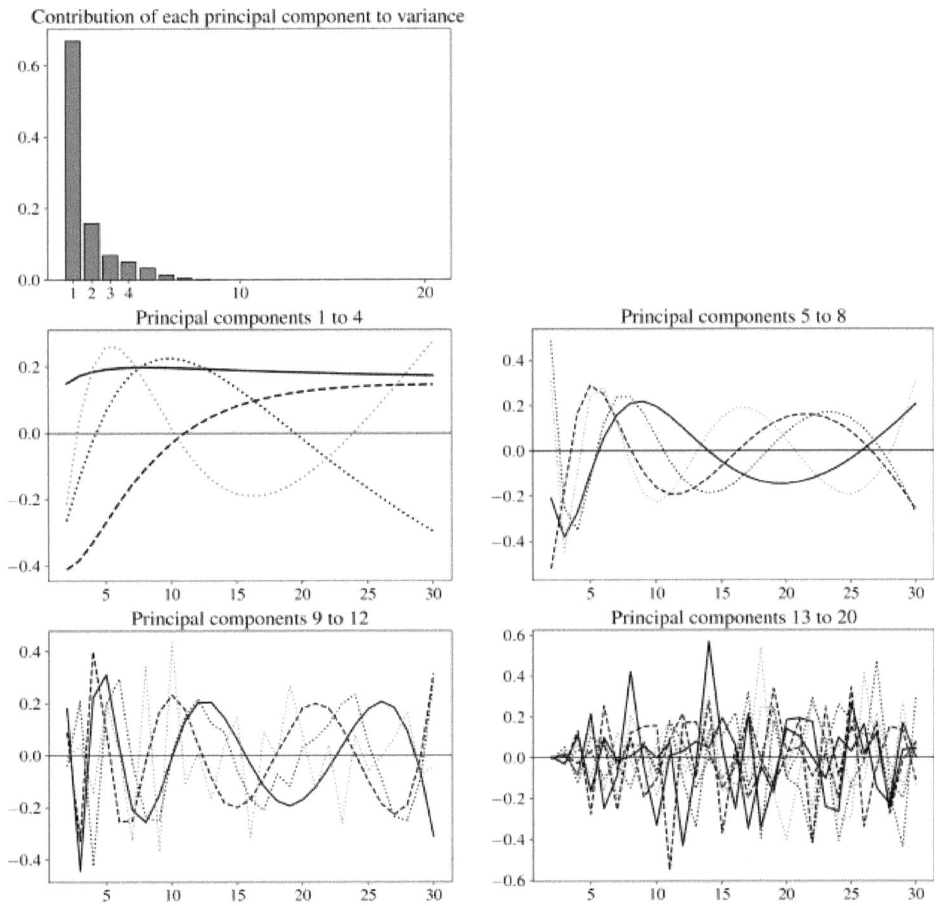

Figure 10.1: Principal components analysis, US Treasury yields, 1984–2020.

As the data above shows, most of the variation can be explained by a small number, say $P < I$, of the principal components with the largest eigenvalues. In fixed income, $P = 2$ principal components are usually enough to explain 90 to 95% of the variation of the yield curve $\mathbf{r}(\omega)$. Often the eigenvector $\mathbf{w}^1$ of the first principal component $x_1(\omega)$ is proportional to the vector of ones. We say then that the first component represents a *parallel yield curve* shift since all rates move by the same amount. This property carries over to a bank portfolio. The return $r^{\mathrm{p}}(\omega)$ of a portfolio of fixed income securities, loans, and deposits will also be (mostly) exposed to $x_1(\omega)$ and $x_2(\omega)$.

The risk management technique to hedge that risk is often called *immunization*. In that technique, P independent interest rate derivatives are chosen to hedge the portfolio. When $P = 1$ is chosen, and the first principal component is a parallel yield curve shift, the technique is called *duration matching*. Some spectacular bankruptcies resulted from hedging only one component of the risk: in 1994, Orange County in California went bankrupt. So did Silicon Valley Bank in 2023. Of course, there were other reasons that contributed to its demise.

10.3.1 ▪ Principal Component Regression

In machine learning one uses the term *ensembling* for combining different ML methods. A good example of ensembling is principal component regression (PCR). The usefulness of PCR is perhaps best illustrated by two examples: *index tracking* and *hedging*.

In **index tracking**, one tries to track the return of an index of I securities by investing in a smaller number P of (possibly different) replicating securities.

The return of the securities that compose the index is denoted by $\mathbf{r}(\omega)$ as before. The problem is, of course, trivial if (i) one knows exactly the composition of the index, and (ii) the component securities are as cheap to buy and sell as the replicating securities. In the other case, one can use principal component regression, which consists of the first two operations of the list below:

- PCA of the returns $\mathbf{r}(\omega)$ resulting in I principal components $\{x_{p,t}\}$
- regression of the index return r_p on the P first principal components
- construction of the synthetic portfolio from replicating securities.

We call *synthetic portfolio* a portfolio with returns equal to $a^* + \sum_{p=1}^{P} b_p^* x_{p,t}$, where $a^*, \mathbf{b}^*$ solve the following minimization problem:

$$\min_{a,\mathbf{b}} \sum_{t=1}^{T} \left(r_{p,t} - a - \sum_{p=1}^{P} b_p x_{p,t} \right)^2$$

We say that the synthetic portfolio either tracks or *replicates* the index.

The problem of **hedging** a portfolio with return r_p is exactly the opposite of the problem of tracking an index r_p. The weights of the hedging portfolio will be the negative of the weights of the replicating portfolio. For fixed income portfolios, the typical hedging instruments are interest rate swaps, whose statistical properties are equivalent to the statistical properties of $\mathbf{r}$. PCR is then useful for two main reasons:

1. While the identity of the underlying factors $\mathbf{r}$ is known (they are interest rates with different terms), the true functional form of r_p, as a function of $\mathbf{r}$, is often complicated and nonlinear and best recovered by statistical analysis.
2. It is difficult to find I cheap hedging instruments and easier to find $P \ll I$ cheap hedging instruments.

10.4 ▪ Clustering

> Cluster analysis, also called *data segmentation*, has a variety of goals. All relate to grouping or segmenting a collection of objects into subsets or "clusters," such that those within each cluster are more closely related to one another than objects assigned to different clusters. An object can be described by a set of measurements or by its relation to other objects. In addition, the goal is sometimes to arrange the clusters into a natural hierarchy. This involves successively grouping the clusters themselves so that at each level of the hierarchy, clusters within the same group are more similar to each other than those in different groups.
>
> (Hastie, Tibshirani, and Friedman [24])

See Figure 10.2. Cluster analysis is also used to form descriptive statistics for each cluster.

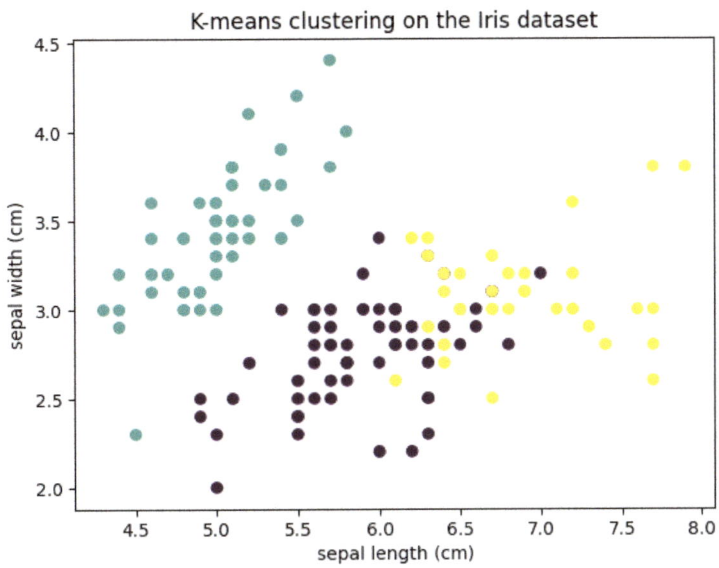

Figure 10.2: K-means clustering on the Iris dataset. Points of the same color belong to the same cluster [29].

Clustering has been applied to many areas of science. There are different types of approaches to cluster data:

- in **hierarchical clustering**, the algorithm tries to build a hierarchy of clusters, following either a bottom-up or a top-down approach
- in **centroid-based clustering**, each cluster is represented by its mean
- in **density-based clustering**, each cluster is defined as areas of higher density
- in **distribution-based clustering**, one assumes the data has been generated from several distributions.

In finance, applications include credit scoring, stock prediction, portfolio selection, and trading strategies analysis.

10.4.1 ▪ Mathematical Setup

Each observation is uniquely labeled by an integer $t \in \{1, \dots, T\}$. A prespecified number of clusters K is postulated, and each one is labeled by an integer $k \in \{1, \dots, K\}$. Each observation

$\mathbf{x}_t \in \mathbb{R}^I$ is assigned to one and only one cluster, in the form of an *encoder*, i.e., a function C such that a cluster $k = C(t)$ for each observation t. One seeks the encoder C^* that achieves the best goal. Central to cluster analysis is the notion of *degree of similarity* (or dissimilarity) between the individual objects being clustered. For instance, squared Euclidean distance

$$d(\mathbf{x}_t, \mathbf{x}_{t'}) = \sum_{i=1}^{I} (x_{t,i} - x_{t',i})^2 = ||\mathbf{x}_t - \mathbf{x}_{t'}||^2$$

is chosen as the dissimilarity measure. A natural loss function to minimize is then

$$W(C) = \frac{1}{2} \sum_{k=1}^{K} \sum_{C(t)=k} \sum_{C(t')=k} ||\mathbf{x}_t - \mathbf{x}_{t'}||^2 \tag{10.6}$$

It measures the distance between each observation in the same cluster.

The most popular algorithm in clustering is probably the K-means algorithm (Lloyd [27], Forgy [28]). However, it is not necessarily globally convergent [24]. We present it here because it can be viewed as a special case of *expectation-maximization* (EM) algorithms, to which several algorithms of this book belong.

10.4.2 ▪ K-Means Algorithm

The K-means algorithm is one of the most popular iterative descent centroid-based clustering methods. It is intended for situations in which all variables are of the quantitative type, and squared Euclidean distance is chosen as the dissimilarity measure. Let $N_k = \sum_{t=1}^{T} 1\{C(t) = k\}$, i.e., the number of points in each cluster, and let $\mathbf{m}_k$ be the mean vector of cluster k, i.e.,

$$\mathbf{m}_k = \sum_{C(t)=k} \mathbf{x}_t / N_k$$

The K-means algorithm is based on the following theorem, which states that the dissimilarity measure (for each cluster) can be computed by taking the sum of the square of the distances between each observation and the mean vector of the cluster.

Theorem 10.5.

$$W(C) = \sum_{k=1}^{K} N_k \sum_{C(t)=k} ||\mathbf{x}_t - \mathbf{m}_k||^2 \tag{10.7}$$

To better understand this theorem, let us focus on one term of the sum in (10.6). In other terms, there is only one cluster, i.e., $k = K = 1$. Let us assume there are only two observations, i.e., $T = 2$. The loss function can then be rewritten as

$$\begin{aligned} W(C) &= \frac{1}{2} \sum_{t=1}^{2} \sum_{t'=1}^{2} ||\mathbf{x}_t - \mathbf{x}_{t'}||^2 = \mathbf{x}_1'\mathbf{x}_1 + \mathbf{x}_2'\mathbf{x}_2 - 2\mathbf{x}_1'\mathbf{x}_2 \\ &= 2\mathbf{x}_1'\mathbf{x}_1 + 2\mathbf{x}_2'\mathbf{x}_2 - (\mathbf{x}_1'\mathbf{x}_1 - \mathbf{x}_1'\mathbf{x}_2 - \mathbf{x}_2'\mathbf{x}_1 - \mathbf{x}_2'\mathbf{x}_2) \\ &= 2\sum_{t=1}^{2} ||\mathbf{x}_t||^2 - \sum_{t=1}^{2}\sum_{t'=1}^{2} \mathbf{x}_t'\mathbf{x}_{t'} \\ &= 2\sum_{t=1}^{2} \left(||\mathbf{x}_t||^2 - ||\mathbf{m}_k||^2\right) \end{aligned}$$

Generalizing this to more observations, we see that $W(C) = N_k \sum_{C(t)=k} \left(||\mathbf{x}_t||^2 - ||\mathbf{m}_k||^2\right)$. You will show in the exercise section that this term can be transformed into the right-hand side of (10.7). This will prove the theorem.

We will not discuss the optimal choice of the number of clusters K, nor will we describe the convergence criterion. As usual, the algorithm starts with the (random) choice of an initial encoder $C^{(0)}$. Once it is given, the algorithm alternates between

- an *expectation* step, where the optimal centers $\mathbf{m}_1, \ldots, \mathbf{m}_K$ of the clusters are found
- a *minimization* step, where the new encoder $C^{(1)}$ is found, i.e., clusters are modified.

The expectation step deserves its name for the following reason. Thanks to the theorem, clearly $\mathbf{m}_k$ (the cluster mean) minimizes $W(C^{(0)})$ for a given encoder $C^{(0)}$. The minimization step corrects the random selection of the initial encoder by assigning each observation to the cluster to which center it is the closest, thus defining a new encoder $C^{(1)}$.

ALGORITHM 10.1. K-means algorithm.

1. Choose a number of clusters K.
2. Set $n = 0$. Choose an initial encoder $C^{(0)}$.
3. Repeat
   For $k = 1$ to K, set
   $\mathbf{m}_k = \frac{1}{N_k} \sum_{C^{(n)}(t)=k} \mathbf{x}_t$
   For $t = 1$ to T, set
   $C^{(n+1)}(t) = \arg\min_k (||\mathbf{x}_t - \mathbf{m}_k||)$
   If converged, exit loop.
   Set $n \leftarrow n + 1$
4. End Repeat

Example 10.6. Suppose we want to classify 25 stocks into three clusters. Each stock has five time observations. For instance, the stocks return at different times. This gives us the constants $K = 3$, $T = 25$, and $I = 5$. In the second step, the algorithm will calculate

$$\mathbf{m}_1 = \frac{1}{N_1} \sum_{C(t)=1} \mathbf{x}_t$$

$$\mathbf{m}_2 = \frac{1}{N_2} \sum_{C(t)=2} \mathbf{x}_t$$

$$\mathbf{m}_3 = \frac{1}{N_3} \sum_{C(t)=3} \mathbf{x}_t$$

In the third step, we assign each of the 25 stocks $\{t = 1, \ldots, T\}$ to the cluster k whose center $\mathbf{m}_k$ is the closest to $\mathbf{x}_t$.

10.5 ▪ To Probe Further

The recent survey by Tang, Tian, and Wu [30] shows that the most commonly used techniques for clustering in finance are hierarchical clustering and centroid-based clustering. See Tripathi et al. [31] for a recent application of clustering to credit scoring. In Dose and Cincotti [32], clustering is applied to index tracking. Pattarin, Paterlin, and Minerva [33] use clustering on financial time series to classify mutual funds better than the usual industry classification. These authors use clustering in conjunction with other techniques. Fricke [34] examines correlations and clustering of trading strategies. This information is important for designing better strategies in illiquid markets (see Chapter 19). At this point, the academic finance literature does not contain too many examples of successful experiments of clustering applied to empirical asset pricing, except in conjunction with other techniques. Patton and Weller [35] use clustering together with OLS to estimate linear factor models.

Exercises

10.1. (Hard) **General Principle of Nonlinear Estimation** (Theorem 10.1)
You will prove that the conditional expectation is the best estimator (in the L^2 sense) of a random variable, i.e., prove that (10.2) holds:

$$E[\tilde{x}_2|\tilde{x}_1] = \arg\min_{\hat{x}_2 \in \mathcal{M}} E[(\tilde{x}_2 - \hat{x}_2)^2] \tag{10.8}$$

where $\mathcal{M}$ is the vector space of random variables that are functions of the random variable $\tilde{x}_1$, have finite second moments, and are closed under that metric (see Hoel, Port, and Stone [36, Chapter 6] for more details).

(a) Show that $\tilde{x}_2^*$ solves (10.8) if $\tilde{x}_2^*$ is the projection of $\tilde{x}_2$ on the space $\mathcal{M}$

$$E[(\tilde{x}_2 - \tilde{x}_2^*)\tilde{z}] = 0 \quad \forall \tilde{z} \in \mathcal{M}. \tag{10.9}$$

(b) Show that $E[\tilde{x}_2^*|\tilde{x}_1]$ solves (10.9).

In other words, $E[\tilde{x}_2^*|\tilde{x}_1]$ is the projection of $\tilde{x}_2^*$ on the space $\mathcal{M}$.

Hint for part (a): use a variational approach. Perturb $\tilde{x}_2^*$ into $\tilde{x}_2^* + a\tilde{z}$ for any $\tilde{z} \in \mathcal{M}$. The right-hand side of (10.8) is now a function of $a \in \mathbb{R}$, which you can differentiate.

10.2. (Hard) You will prove the centroid equality of clustering (10.7) by taking the following steps. First, suppose that the number of clusters is $k = K = 1$. Then

(a) show that $W(C) = N_k \sum_{C(t)=k} \left(||\mathbf{x}_t||^2 - ||\mathbf{m}_k||^2\right)$ by following the steps in this book;

(b) show that

$$\sum_{C(t)=k} ||\mathbf{x}_t - \mathbf{m}_k||^2 = \sum_{C(t)=k} \left(||\mathbf{x}_t||^2 - ||\mathbf{m}_k||^2\right) \tag{10.10}$$

The last step can be recognized as different formulas for the empirical variance.

10.3. In this problem you will cluster stocks according to their stock price at different periods.

The data is given in the spreadsheet ch10exercise3data.csv in the supplementary material.[31] We look for three clusters, i.e., $K = 3$.
We have $T = 25$ (25 portfolios) and $I = 5$ (five time observations: Sept, Oct, Nov, Dec 2020 and Jan 2021).

[31] https://bookstore.siam.org/ot195/bonus

10.4. Use the data from Exercise 6.5. Perform a principal component analysis of the returns of the six stocks $\{r_{i,t}\}$ defined in Exercise 9.5: Google, Walmart, Apple, GM, Intel, Pfizer.

(a) What are the first two principal components $\mathbf{w}^1$ and $\mathbf{w}^2$ (seen as row vectors)?

(b) Construct two portfolios $x_{1,t} = \mathbf{w}^1\mathbf{r}_t$ and $x_{2,t} = \mathbf{w}^2\mathbf{r}_t$

(c) Regress the S&P 500 return on the two portfolios above. Do you get a better or worse R-squared than by regressing directly the S&P 500 return on the six stocks above?

Use for the training period April 1st to August 1st, 2021, and for the test period August 2nd to October 1st, 2021.

10.5. (Medium) Fix x_1. Provide a supervised learning methodology to construct the empirical estimator of $P(\tilde{x}_2 = x_2|\tilde{x}_1 = x_1)$. Hint: Recall that for any event A, $P(A) = E[1[A]]$, where the *indicator function* applied to A, namely, $1[A]$, takes the value 1 if A is true and 0 otherwise. Warning: this is not an efficient method.

Chapter 11

Deep Feedforward Neural Networks

Deep feedforward networks, also called **feedforward neural networks** or **multilayer perceptrons** (MLPs), are the most novel technique in supervised learning. They belong to the family of nonlinear regression methods, or nonlinear function approximation. A feedforward network defines a mapping $\mathbf{f}$ between examples $\mathbf{x} \in \mathbb{R}^I$ and labels $\mathbf{y} \in \mathbb{R}^J$. It defines the mapping by selecting a parameter vector $\boldsymbol{\theta} \in \mathbb{R}^K$ that results in the best approximation $\mathbf{f}$ of the true function $\mathbf{f}^*$:

$$\mathbf{y} = \mathbf{f}^*(\mathbf{x}) + \mathbf{u} \tag{11.1}$$

where $\mathbf{u}$ is the intrinsic error. For instance, the linear regression model in (6.1) with $J = 1$, namely, $y_t = \mathbf{x}_t'\boldsymbol{\beta} + u_t$, is a special case where the parameter $\boldsymbol{\theta}$ is called $\mathbf{b}$ and where

$$f(\mathbf{x}; \boldsymbol{\theta}) = \mathbf{x}'\mathbf{b} \tag{11.2}$$

$$f^*(\mathbf{x}) = \mathbf{x}'\boldsymbol{\beta} \tag{11.3}$$

The methods are called **feedforward** because information flows through the function being evaluated at $\mathbf{x}$ through the intermediate computations used to define $\mathbf{f}$ and finally to the output $\mathbf{y}$. When feedforward neural networks are extended to include feedback connections, they are called **recurrent neural networks**, as we shall see in Chapter 12.

If a neural network has more than one **layer**, it is usually called a **deep neural network**. What is a layer? It is simply a function. For instance, a three-layer network is defined by three functions: $\mathbf{f}^{(1)}$ for the first layer, $\mathbf{f}^{(2)}$ for the second one, and $\mathbf{f}^{(3)}$ for the third one. The full network is defined by the function $\mathbf{f}$ defined by the composition of the layers:

$$\mathbf{f}(\mathbf{x}) = \mathbf{f}^{(3)}(\mathbf{f}^{(2)}(\mathbf{f}^{(1)}(\mathbf{x})))$$

The number of layers is called the **depth** of the neural network. The layers whose input is not (directly) visible are called the **hidden layers**.

This chapter starts with a classical example, which shows that "linear functions cannot approximate everything," and thus one needs to include some nonlinearity in function approximation. We then move to a description of the standard architecture of feedforward neural networks, for which a MATLAB code was given in Exercise 2.2. We then motivate the architecture and suggest enhancements.

11.1 ▪ Example: Learning XOR by Linear Functions

The XOR function takes two binary variables, x_1 and x_2, and outputs the result $y = f^*(\mathbf{x})$ according to the following table:

	$x_2 = 0$	$x_2 = 1$
$x_1 = 0$	$y = 0$	$y = 0$
$x_1 = 1$	$y = 0$	$y = 1$

Suppose there is no noise. The MSE loss function is then

$$J(\theta) = \frac{1}{4}\sum_{x_1=0}^{1}\sum_{x_2=0}^{1}(f^*(\mathbf{x}) - f(\mathbf{x};\boldsymbol{\theta}))^2$$

Let us take a linear function. The parameter $\boldsymbol{\theta} = (\mathbf{w}, b)$, and

$$f(\mathbf{x}; \mathbf{w}, b) = \mathbf{x}'\mathbf{w} + b$$

After solving the normal equations, we obtain

$$\mathbf{w}^* = 0$$
$$b^* = 1/2$$

which gives an obviously poor approximation. One may be tempted to believe that this example definitely discards models that are linear in the parameters. The reality is more subtle, as you will see in Exercise 11.1. Another example of the need for nonlinear functions of the parameters is described in Example 6.10.

11.2 ▪ Standard Architecture

In what follows, we fully describe the deep feedforward network, whose MATLAB code is in Exercise 2.2. We defer the discussion of some variations around it, as well as the choice of the hyperparameters, to another section.

As elsewhere in this book, we suppose there are T examples in the training dataset $\{\mathbf{x}_1, \ldots, \mathbf{x}_t, \ldots, \mathbf{x}_T\}$. Returning to the notation $J(\boldsymbol{\theta})$ for the MSE as a function of the parameters, the training problem is to minimize $J(\boldsymbol{\theta})$:

$$J(\boldsymbol{\theta}) = \sum_{t=1}^{T} ||\mathbf{y}_t - \mathbf{f}(\mathbf{x}_t; \boldsymbol{\theta})||^2 \tag{11.4}$$

The feedforward pass calculates the function $\mathbf{f}(\mathbf{x}_t; \boldsymbol{\theta})$ for a given input $\mathbf{x}_t$ while the backward pass calculates the gradient of a *reduced MSE* $||\mathbf{y}_t - \mathbf{f}(\mathbf{x}_t; \boldsymbol{\theta})||^2$ for some t arbitrarily chosen.

11.2.1 ▪ Feedforward Pass

We denote the number of layers by L. Each layer will be identified by the superscript l, where $l = 1$ is the input layer and $l = L$ is the output layer. Each layer is composed of n_l **neurons**. Thus the full network is a function that maps n_1 inputs to n_L outputs:

$$\mathbf{f}(.; \boldsymbol{\theta}) : \mathbb{R}^{n_1} \to \mathbb{R}^{n_L}$$

Each neuron j in layer l takes n_{l-1} inputs and produces one output. Each neuron is thus a function $f_j^{(l)}$:

$$f_j^{(l)}(.; \boldsymbol{\theta}) : \mathbb{R}^{n_{l-1}} \to \mathbb{R}$$

Denote by $a_j^{(l)}$ the output of each neuron:

$$a_j^{(l)} = f_j^{(l)}(\mathbf{a}^{(l-1)}; \theta) \tag{11.5}$$

with

$$\begin{aligned} \mathbf{a}^{(1)} &= \mathbf{x} \\ \mathbf{a}^{(L)} &= \mathbf{y} \end{aligned}$$

Layer l is thus a vector-valued function $\mathbf{f}^{(l)}$ that concatenates the action of each of its neurons:

$$\mathbf{a}^l = \mathbf{f}^{(l)}(\mathbf{a}^{l-1}) = \begin{bmatrix} f_1^{(l)}(\mathbf{a}^{l-1}) \\ \vdots \\ f_{n_l}^{(l)}(\mathbf{a}^{l-1}) \end{bmatrix}$$

Since the layer is feedforward, the output of layer $l-1$, namely, $\mathbf{a}^{l-1}$, is the input of layer l. The full network is thus given by the composition of functions:

$$\mathbf{f} = \mathbf{f}^{(L)} \circ \mathbf{f}^{(L-1)} \circ \cdots \circ \mathbf{f}^{(2)}$$

Each layer is composed of two steps. First, it transforms the input data linearly by using a **weight** matrix $\mathbf{W}^{(l)}$ and a **bias** vector $\boldsymbol{b}^{(l)}$:

$$\mathbf{z}^{(l)} = \mathbf{W}^{(l)}\mathbf{a}^{(l-1)} + \boldsymbol{b}^{(l)}$$

Second, the output of each neuron is given by an $\mathbb{R} \to \mathbb{R}$ function σ:

$$a_j^{(l)} = \sigma(z_j^{(l)})$$

where σ is called the **activation** function. In this particular architecture, we choose the *sigmoid*:

$$\sigma(z) = \frac{1}{1+e^{-z}} \tag{11.6}$$

Thus each neuron j in layer l is described by

$$f_j^{(l)}(\mathbf{a}^{(l-1)}; \mathbf{W}^{(l)}, \boldsymbol{b}^{(l)}) = \sigma((\mathbf{W}^{(l)}\mathbf{a}^{(l-1)} + \boldsymbol{b}^{(l)})_j)$$

It is practical to concatenate the output of each layer into a vector-valued function, whose dimension can change depending on the layer:

$$\boldsymbol{\sigma}^{(l)}(\mathbf{z}) = \begin{bmatrix} \sigma(z_1) \\ \vdots \\ \sigma(z_{n_l}) \end{bmatrix}$$

We can then compress notation into

$$\begin{aligned} \mathbf{f}^{(l)}(\mathbf{a}^{(l-1)}; \mathbf{W}^{(l)}, \boldsymbol{b}^{(l)}) &= \boldsymbol{\sigma}^{(l)}(\mathbf{z}^{(l)}) \\ &= \boldsymbol{\sigma}^{(l)}(\mathbf{W}^{(l)}\mathbf{a}^{(l-1)} + \boldsymbol{b}^{(l)}) \end{aligned}$$

Recalling (11.5), the output $\mathbf{a}^l$ of the lth layer can be expressed as a function of the output $\mathbf{a}^{(l-1)}$ of the $(l-1)$th layer. Thus the full network is

$$\mathbf{a}^1 = \mathbf{x} \tag{11.7}$$

$$\mathbf{a}^l = \boldsymbol{\sigma}^{(l)}(\mathbf{W}^{(l)}\mathbf{a}^{(l-1)} + \boldsymbol{b}^{(l)}) \qquad l \in [2, L] \tag{11.8}$$

$$\mathbf{f}(\mathbf{x}; \boldsymbol{\theta}) = \mathbf{a}^L \tag{11.9}$$

The vector-valued sigmoid function $\boldsymbol{\sigma}^{(l)}$ can be implemented as follows in MATLAB:

```
function y=activate(x,W,b)
y=1./(1+exp(-(W*x+b)));
```

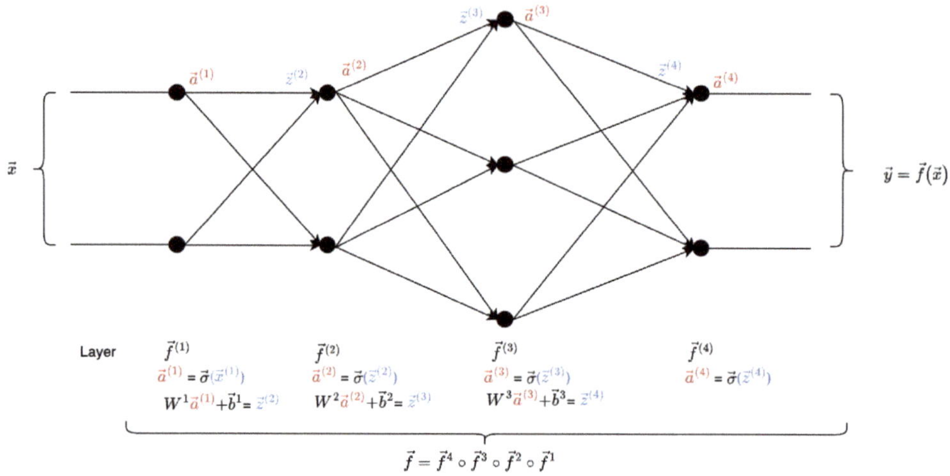

Figure 11.1: A four-layer (feedforward) neural network.

Example 11.1. Suppose we have an $L = 4$ layer network (see Figure 11.1) with

- $n_1 = 2$
- $n_2 = 2$
- $n_3 = 3$
- $n_4 = 2$.

The feedforward calculation realizes

$$\begin{aligned}
\mathbf{a}^2 &= \boldsymbol{\sigma}^{(2)}(\mathbf{z}^{(2)}) = \boldsymbol{\sigma}^{(2)}(\mathbf{W}^{(2)}\mathbf{x} + \mathbf{b}^{(2)}) \\
\mathbf{a}^3 &= \boldsymbol{\sigma}^{(3)}(\mathbf{z}^{(3)}) = \boldsymbol{\sigma}^{(3)}(\mathbf{W}^{(3)}\mathbf{a}^2 + \mathbf{b}^{(3)}) \\
\mathbf{a}^4 &= \boldsymbol{\sigma}^{(4)}(\mathbf{z}^{(4)}) = \boldsymbol{\sigma}^{(4)}(\mathbf{W}^{(4)}\mathbf{a}^3 + \mathbf{b}^{(4)})
\end{aligned}$$

Since $\mathbf{a}^4 = \mathbf{f}(\mathbf{x};\theta)$, the full network realizes the function

$$\mathbf{f}(\mathbf{x};\boldsymbol{\theta}) = \underbrace{\boldsymbol{\sigma}^{(4)}(\mathbf{W}^{(4)}\underbrace{\boldsymbol{\sigma}^{(3)}(\mathbf{W}^{(3)}\underbrace{\boldsymbol{\sigma}^{(2)}(\mathbf{W}^{(2)}\mathbf{x} + \mathbf{b}^{(2)})}_{\mathbf{f}_2(\mathbf{x})} + \mathbf{b}^{(3)})}_{\mathbf{f}_3(\mathbf{f}_2(\mathbf{x}))} + \mathbf{b}^{(4)})}_{\mathbf{f}_4(\mathbf{f}_3(\mathbf{f}_2(\mathbf{x})))}$$

We have the following dimensions of the parameters:

	$l=2$	$l=3$	$l=4$
$\mathbf{W}^{(l)}$	$\mathbb{R}^{2,2}$	$\mathbb{R}^{3,2}$	$\mathbb{R}^{2,2}$
$\mathbf{b}^{(l)}$	$\mathbb{R}^{2}$	$\mathbb{R}^{3}$	$\mathbb{R}^{2}$

So the parameter $\boldsymbol{\theta} = \{\mathbf{W}^{(2)}, \mathbf{W}^{(3)}, \mathbf{W}^{(4)}, \mathbf{b}^{(2)}, \mathbf{b}^{(3)}, \mathbf{b}^{(4)}\}$, and we have a total of $K = 21$ parameters.

The MATLAB code for this example is

```
a2=activate(x,W2,b2);
a3=activate(a2,W3,b3);
a4=activate(a3,W4,b4);
```

Observe an advantage of MATLAB: the dimensions do not need to be coded.

Summary

The output of the forward pass in the example is a2, a3, a4.

11.2.2 ▪ Stochastic Gradient Algorithm

When T is large, computing the gradient vector $\nabla J(\boldsymbol{\theta})$ for each element in the sum in equation (11.4) is prohibitively expensive. A much cheaper alternative is to replace the mean of the gradients over the training points by the gradient at a single randomly chosen training point. We then perform the gradient algorithm on this reduced cost function, in this example for N iterations.

ALGORITHM 11.1. Stochastic gradient algorithm.

1. Set $n=0$, $\boldsymbol{\theta}^{(0)} =$ any value, learning rate η, N
2. Repeat
   Choose an integer t at random from $\{1,\ldots,T\}$
   Set $\boldsymbol{\theta}^{(n+1)} = \boldsymbol{\theta}^{(n)} - \eta\nabla_{\boldsymbol{\theta}^{(n)}}||\mathbf{y}_t - \mathbf{f}(\mathbf{x}_t;\boldsymbol{\theta}^{(n)})||^2$
   $n \longleftarrow n+1$
3. Until $n=N$
4 Output $\boldsymbol{\theta}^{(N)}$

11.2.3 ▪ Backpropagation

Notation: the derivative of the reduced cost function $||\mathbf{y}_t - \mathbf{f}(\mathbf{x}_t;\boldsymbol{\theta})||^2$ with respect to the hidden variables z_j^l (input of the jth neuron in layer l) is denoted by

$$\delta_j^{[l]} = \frac{\partial||\mathbf{y}_t - \mathbf{f}(\mathbf{x}_t;\boldsymbol{\theta})||^2}{\partial z_j^l}$$

If $\mathbf{x},\mathbf{y} \in \mathbb{R}^n$, we denote by $\mathbf{x}\circ\mathbf{y} \in\mathbb{R}^n$ the Hadamard or componentwise product, that is,

$$(\mathbf{x}\circ\mathbf{y})_i = x_i y_i$$

Backpropagation is a computationally economical method to recursively calculate the gradient $\nabla_{\boldsymbol{\theta}^{(n)}}||\mathbf{y}_t - \mathbf{f}(\mathbf{x}_t;\boldsymbol{\theta}^{(n)})||^2$. By the chain rule of calculus, the derivatives at the lth layer will depend on the derivatives of the $(l+1)$th layer. The algorithm first calculates $\boldsymbol{\delta}^{[L]}$, and then $\boldsymbol{\delta}^{[L-1]}$, all the way to $\boldsymbol{\delta}^{[2]}$. Applying the chain rule results in the following lemma.

Lemma 11.2.

$$\begin{aligned}
\boldsymbol{\delta}^{[L]} &= (\boldsymbol{\sigma}^{(L)})'(\mathbf{z}^L) \circ (\mathbf{a}^L - \mathbf{y}) \\
\boldsymbol{\delta}^{[l]} &= (\boldsymbol{\sigma}^{(l)})'(\mathbf{z}^l) \circ (\mathbf{W}^{l+1})^t \boldsymbol{\delta}^{[l+1]} \quad \text{for } l < L \\
\frac{\partial ||\mathbf{y}^t - \mathbf{f}(\mathbf{x}^t;\boldsymbol{\theta})||^2}{\partial \mathbf{b}^l} &= \boldsymbol{\delta}^{[l]} \\
\frac{\partial ||\mathbf{y}^t - \mathbf{f}(\mathbf{x}^t;\boldsymbol{\theta})||^2}{\partial W^l_{j,k}} &= \delta^{[l]}_j a^{l-1}_k
\end{aligned}$$

Proof. This is Lemma 5.1 in Higham and Higham [37]. You will be asked to prove it in Exercise 11.5.

Observe that, for the sigmoid (11.6),

$$\sigma'(x) = \sigma(x)(1 - \sigma(x))$$

Denote by $\mathbf{D}^l$ the diagonal matrix with $\mathbf{D}^l(i,i) = \sigma'(z^l_i)$. We can thus expand

$$\boldsymbol{\delta}^{[l]} = \mathbf{D}^l(\mathbf{W}^{l+1})^t \mathbf{D}^{l+1}(\mathbf{W}^{l+2})^t \ldots \mathbf{D}^L(\mathbf{a}^L - \mathbf{y})$$

The full backpropagation algorithm is then as follows.

ALGORITHM 11.2. Backpropagation.
$\boldsymbol{\delta}^{[L]} = \mathbf{D}^L(\mathbf{a}^L - \mathbf{y})$
For $l = L - 1$ downto 2
  $\boldsymbol{\delta}^{[l]} = \mathbf{D}^l(\mathbf{W}^{l+1})^t \boldsymbol{\delta}^{[l+1]}$
end

Example 11.3. The MATLAB code for our example is (again, for $L = 4$)

```
delta4=a4.*(1-a4).*(a4-y(:,t));
delta3=a3.*(1-a3).*(transpose(W4)*delta4);
delta2=a2.*(1-a2).*(transpose(W3)*delta3);
W2=W2-eta*delta2*x';
W3=W3-eta*delta3*a2';
W4=W4-eta*delta4*a3';
b2=b2-eta*delta2;
b3=b3-eta*delta3;
b4=b4-eta*delta4;
```

Summary

The output of the backpropagation algorithm in our example is the updated parameter $\boldsymbol{\theta}^{(n+1)}$.

11.2.4 ▪ Full Algorithm: Example

Recall that in our example $\mathbf{x}_t$ has $I = 2$ dimensions:

$$\mathbf{x}_t = \begin{bmatrix} x_{1,t} \\ x_{2,t} \end{bmatrix}$$

while y_t has one dimension. Modulo inserting the forward pass and the backpropagation, and

Input/Output functions, the MATLAB code for our example above, with number of layers $L = 4$, is

```
%Step 1: initialization
    rng(5000);
    W2=0.5*randn(2,2); W3=0.5*randn(3,2); W4=0.5*randn(2,3);
    b2=0.5*randn(2,1); b3=0.5*randn(3,1); b4=0.5*randn(2,1);
    eta=0.05; %learning rate
    Niter=1e6; %number of Stochastic Gradient (SG) iterations
%Step 2: optimization loop
    for counter=1:Niter
        t=randi(T);
        x=[x1(t);x2(t)];
        Feedforward pass (see code above)
        Backpropagation (see code above)
    end
```

Again, the full code is shown in Exercise 2.2.

11.3 ▪ Motivation and Enhancements

We evaluate in turn every part of the architecture above.

11.3.1 ▪ The Activation Function

The initial choice for the activation function was the sigmoid function. The idea was clear: like a neuron in the brain, the sigmoid function caps the signals; the sigmoid outputs values between zero and one. A drawback of this design is that the gradient can become very small when the input of the neuron is very large (in absolute value).

For that reason, most researchers are now applying the rectified linear unit (ReLu) function. See Figure 11.2.

$$\text{Relu}(z) = \max(0, z)$$

The disadvantage of the ReLu function is that it can become unbounded. However, the gradient is more "friendly." When the input z is negative, the neuron is "deactivated," i.e., it

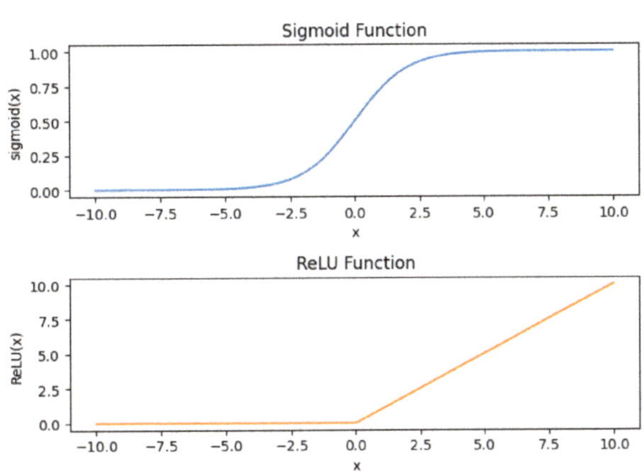

Figure 11.2: ReLu and sigmoid functions.

does not contribute to the gradient of the cost function. When the input z is positive, the neuron behaves like a regular linear function, and its gradient is easy to calculate. One may wonder: do we not lose one part of the input space, i.e., the part where $z < 0$? The obvious answer is to include more neurons (see Exercise 11.3).

Not all layers need to have the same type of activation function. For instance, when using the sigmoid function as a last layer, the output $\hat{y}_t$ is necessarily between zero and one. If this is not the case for the label $\mathbf{y}_t$, then the algorithm will obviously fail to give a good approximation. An obvious solution is for the last layer to have an **identity activation function**, ($\sigma(z) = z$), resulting in

$$\mathbf{f}^{(L)}(\mathbf{a}^{(L-1)}; \mathbf{W}^{(L)}, \boldsymbol{b}^{(L)}) = \mathbf{W}^{(L)}\mathbf{a}^{(L-1)} + \boldsymbol{b}^{(L)}$$

which can take any possible value.

The identity activation function can also be used in the first hidden layer as a preprocessing unit. The idea then is **dimension reduction**, like in PCA: the number of neurons in the second layer (n_2) would be smaller than the number of neurons in the first layer:

$$n_2 \ll n_1$$

so as to extract the most important combination of features. The full network can then be seen as a (nonlinear) principal component regression.

11.3.2 ▪ Number of Hidden Layers

When the dimension of the data $I = 1$, it is clear that a network with one hidden layer and linear output (effecting a summation) can approximate any continuous function, as you will see in Exercise 11.6.

The **universal approximation theorem** (Hornik, Stinchcombe, and White [38]; Cybenko [41]) shows that only one hidden layer is needed to approximate any function of several variables using the sigmoid activation function. Yet, it is more efficient to have more than one hidden layer. Researchers have been trying for a long time to approximate functions by composition of more elementary functions. For instance, the famous "father of modern probability theory," Kolmogorov [39], found that a function of many variables can be built up by composing (and adding) functions of only three variables.[32] His student Arnold [40] proved that it was possible with functions of only two variables.

Unfortunately, none of these results is constructive, as far as we know. This is the reason one uses the stochastic gradient algorithm to build the approximation.

Finding the optimal number of hidden layers for a given task has been and still is an intense area of research, which is beyond the scope of this book. We refer the reader to the book *Deep Learning* by Goodfellow, Bengio, and Courville [8].

11.3.3 ▪ Minimization Algorithm

The main (not completely resolved) question about MLP is the question of **convergence**. Neural networks attempt to solve a (continuous) **global optimization** problem. It is at this point unknown whether there is a polynomial-type algorithm to solve continuous global optimization problems. If there is, this would probably solve the famous long-term conjecture in computer science that **P=NP**, which is (at the time of writing this book) widely believed to be false. Thus, the stochastic gradient algorithm may be stuck at a suboptimal local minimum. This is particularly the case with underfitted models. For overfitted models, we will argue in the next paragraph that the error will converge to zero on the training set. What is more surprising, as we saw in

[32] Kolmogorov considered this work to be "his most technically difficult accomplishment" [42].

Chapter 2, is that several researchers (see [11]) claim that the generalization error is also small for overfitted models.

For **overfitted models**, the local minimum problem (on a single dataset) is indeed not as dire as it may seem. Looking at only one dimension, one may believe that the algorithm will be stuck in a suboptimal local minimum. However, one can escape easily from that predicament if one looks in a different direction. This is illustrated in Figure 11.3: Think of two valleys in a mountain range. The valleys are oriented North-South, and the elevation is higher in the North than in the South. For a particular latitude, there are two local minima of the function. An algorithm that thinks only in terms of latitude (going East or West) will not, from a local minimum, naturally join the other minimum. However, if one adds the possibility of moving longitudinally and chooses the South direction, the valleys will ultimately merge and meet the global minimum, namely, sea level, as Figure 11.3 shows. In this stylized example, the number of parameters is $K = 2$ (the two geographic directions). How does such an error function arise from the data? In the next example, we start with the error function and then derive a model and data that matches Figure 11.3.

Figure 11.3: This picture of two valleys merging into one valley shows that, in one direction, there appear to be two local minimizers, but these points are just the bottom part of the valleys, and so they are not local minimizers. Photo taken at Val d'Anniviers, Switzerland.

Example 11.4.[33] For simplicity, we assume no intrinsic error in our dataset ($y = \hat{y}$) so that our model is exact. The model is voluntarily trivial:

$$y = f^*(x) = x$$

[33]While this example should be skipped at first reading, it will help researchers who want to become more proficient at generating test data.

There will also be no approximation error at the optimal parameter $\boldsymbol{\theta}^* = (b^*, w^*) = (0, 1)$, i.e.,

$$f(x, \boldsymbol{\theta}^*) = f^*(x) \tag{11.10}$$

We restrict the parameter w to the interval $[0, 1]$. In our construction, we want the geometry of the error $J(\boldsymbol{\theta})$ to resemble Figure 11.3, that is, two valleys, which are local optima in the direction b but that gradually decrease following an increased value of w and merge at $\boldsymbol{\theta}^*$. The "bottoms of the valleys" are lines in the (w, b) plane, denoted by $b_1(w)$ and $b_2(w)$. For simplicity, we will assume that the data x_t is uniformly spread over the interval $t \in [-1, 1]$ and that data is plentiful, so we can use a continuum approximation. In other words, our dataset is such that

$$\hat{y}_t = x_t = t$$

We take for the objective function to minimize

$$J(\boldsymbol{\theta}) = \exp\left(\int_{-1}^{1} (\hat{y}_t - f(x_t, \boldsymbol{\theta}))^2 dt\right) \tag{11.11}$$

and decompose it into two terms:

$$J(\boldsymbol{\theta}) = J_{pos}(\boldsymbol{\theta}) J_{neg}(\boldsymbol{\theta}) \tag{11.12}$$

Each term can be thought of as an error which, along the b direction, varies only in a particular valley: $J_{pos}(\boldsymbol{\theta})$ will vary in the "positive b valley," while $J_{neg}(\boldsymbol{\theta})$ will vary in the "negative b valley." Correspondingly, the approximating function f will be split in two parts, one, f_{pos}, acting on positive data, and the other one, f_{neg}, on negative data:

$$f(x, \boldsymbol{\theta}) = f_{neg}(x, \boldsymbol{\theta}) 1[x < 0] + f_{pos}(x, \theta) 1[x \geq 0]$$

Function f_{neg} will be constructed to result in the "positive b valley" error, while f_{pos} will be constructed to result in the "negative b valley" error:

$$J_{pos}(\boldsymbol{\theta}) = \exp\left(\int_{-1}^{0} (t - f_{neg}(t, \boldsymbol{\theta}))^2 dt\right) \quad \text{"positive } b \text{ valley, negative data"} \tag{11.13}$$

$$J_{neg}(\boldsymbol{\theta}) = \exp\left(\int_{0}^{1} (t - f_{pos}(t, \boldsymbol{\theta}))^2 dt\right) \quad \text{"negative } b \text{ valley, positive data"}$$

The "positive b valley" has elevation proportional, for fixed w, to

$$J_{pos}(\boldsymbol{\theta}) = \begin{cases} (b - (1 - w))^2 + 2 - w & b \geq 0 \\ (w - 1)^2 + 2 - w & b < 0 \end{cases} \tag{11.14}$$

Its bottom ($\frac{\partial J_{pos}(\boldsymbol{\theta})}{\partial b} = 0$) is described by the coordinate $b_1(w) = 1 - w \geq 0$. The "negative b valley" has elevation proportional, for fixed w, to

$$J_{neg}(\boldsymbol{0}) - \begin{cases} (w - 1)^2 + 2 - w & b \geq 0 \\ (b - (w - 1))^2 + 2 - w & b < 0 \end{cases} \tag{11.15}$$

Its bottom ($\frac{\partial J_2(\boldsymbol{\theta})}{\partial b} = 0$) is described by the coordinate $b_2(w) = w - 1 \leq 0$. Merging the two parts results in the following elevation for the valley bottoms:

$$J(b_1(w), w) = J(b_2(w), w) = (2 - w)((w - 1)^2 + 2 - w)$$

which decreases in w. The minimum is reached at $J(b^*, w^*) = 1$, which because of (11.11) results in no approximation error, i.e., (11.10) is satisfied. A function f_{neg} that satisfies (11.13) must be of the form

$$\int_{-1}^{0} (t - f_{neg}(t, \boldsymbol{\theta}))^2 dt = g_{neg}(0) - g_{neg}(-1)$$

where $g_{neg}(t)$ is the primitive of $(t - f_{neg}(t, \boldsymbol{\theta}))^2$ and so that $g_{neg}(0) - g_{neg}(-1) = \ln(J_{pos}(\boldsymbol{\theta}))$. Rather than taking g_{neg} to be a linear function of t, which would make for a very dull function f_{neg}, we take, for more flexibility,

$$g_{neg}(t) = \ln\left[(b - (1 - w))^2 (t + 1) + (2 - w))\right] (t + 1) \quad \text{for } b \geq 0$$

Differentiating with respect to t gives our approximating function

$$f_{neg}(t, \boldsymbol{\theta}) = t - \sqrt{\left|\frac{dg_{neg}(t)}{dt}\right|}$$

The approximating function $f_{pos}(t, \boldsymbol{\theta})$ is determined the same way, by setting

$$\int_{0}^{1} (t - f_{pos}(t, \boldsymbol{\theta}))^2 dt = \ln(J_{neg}(\boldsymbol{\theta})) \tag{11.16}$$

You will determine a function f_{pos} in Exercise 11.7 and see in Exercise 11.8 how to build an MLP that implements f. See Table 11.1, which shows the resulting error $J(b, w)$.

	$w = 0.1$	$w = 0.2$	$w = 0.3$	$w = 0.4$	$w = 0.5$	$w = 0.6$	$w = 0.7$	$w = 0.8$	$w = 0.9$	$w = 1$
$b = 1$	5.18	4.49	3.92	3.45	3.06	2.75	2.49	2.28	2.12	2.00
	5.15	4.42	3.81	3.31	2.91	2.57	2.31	2.10	1.93	1.81
	5.18	**4.39**	3.74	3.21	2.78	2.43	2.15	1.93	1.76	1.64
	5.26	4.42	**3.72**	3.16	2.70	2.32	2.03	1.80	1.62	1.49
	5.39	4.49	3.74	**3.14**	2.64	2.25	1.93	1.69	1.50	1.36
	5.58	4.61	3.81	3.16	**2.63**	2.20	1.86	1.60	1.40	1.25
	5.83	4.78	3.92	3.21	2.64	**2.18**	1.82	1.54	1.32	1.16
	6.12	5.00	4.07	3.31	2.70	2.20	**1.81**	1.50	1.27	1.09
	6.48	5.27	4.27	3.45	2.78	2.25	1.82	**1.49**	1.23	1.04
	6.88	5.59	4.51	3.63	2.91	2.32	1.86	1.50	**1.22**	1.01
$b = 0$	7.34	5.95	4.80	3.84	3.06	2.43	1.93	1.54	1.23	**1.00**
	6.88	5.59	4.51	3.63	2.91	2.32	1.86	1.50	**1.22**	1.01
	6.48	5.27	4.27	3.45	2.78	2.25	1.82	**1.49**	1.23	1.04
	6.12	5.00	4.07	3.31	2.70	2.20	**1.81**	1.50	1.27	1.09
	5.83	4.78	3.92	3.21	2.64	**2.18**	1.82	1.54	1.32	1.16
	5.58	4.61	3.81	3.16	**2.63**	2.20	1.86	1.60	1.40	1.25
	5.39	4.49	3.74	**3.14**	2.64	2.25	1.93	1.69	1.50	1.36
	5.26	4.42	**3.72**	3.16	2.70	2.32	2.03	1.80	1.62	1.49
	5.18	**4.39**	3.74	3.21	2.78	2.43	2.15	1.93	1.76	1.64
	5.15	4.42	3.81	3.31	2.91	2.57	2.31	2.10	1.93	1.81
$b = -1$	5.18	4.49	3.92	3.45	3.06	2.75	2.49	2.28	2.12	2.00

Table 11.1: Error $J(b, w)$ as a function of b and w. The valley bottoms are denoted in bold. Check the correspondence with Figure 11.3.

Other algorithms

One may think that moving to second-order methods (Newton's method) could help. However, MLP is a method for high-dimensional problems. As alluded to before (see Chapter 10), there are many more saddle points in high dimensions than minima. Since Newton's method is more likely to be trapped in minima, most researchers prefer to use a first-order method. Different variations of the stochastic gradient algorithm exist that use **bagging** (taking several samples at each iteration rather than a single one), or so-called momentum. The *Adam algorithm* can be seen as a nonlinear variation of the Cochrane–Orcutt algorithm for GLS, i.e., implementing **boosting**. We refer the reader to Goodfellow, Bengio, and Courville [8] for this rapidly moving area of research.

Application to Finance

Many models in asset management use a small number of observations (say $T < 120$, i.e., monthly for the last 10 years). The reason is that information becomes obsolete quickly. Many practitioners also prefer to work on robust, underfitted models, that is, a small number of parameters. The reason is that asset managers can better explain to their clients why they make a particular investment decision. In this environment, the stochastic gradient algorithm has two disadvantages:

- The speed gain (of the order of T) compared to the deterministic gradient is barely perceptible; an algorithm that runs in 1 second is barely more useful than an algorithm that runs in 2 minutes, at least in comparison to the weeks of work it takes for analysts to agree on a model.
- The algorithm may fail to find the minimum, even if the problem turns out to be convex.

If the modeler prefers an exactly fitted model, and the model error is known a priori to have no isolated local minima, then classical optimization algorithms (such as those presented in Chapter 9) should be preferred to neural networks.

11.3.4 ▪ Hyperparameters

Besides the structure of the model, the standard MLP has three different types of hyperparameters:

- the learning rate
- the initial (random) values of the parameters
- the termination criterion.

In convex optimization, we saw that the **learning rate** should not be too large, so as not to overshoot a local minimum. When there are several isolated local minima, however, the thinking is different: in order not to be trapped in a suboptimal local minimum, the learning rate should be larger, so that more regions are surveyed by the algorithm. The algorithm then resembles simulated annealing. A popular technique is to use an adaptive learning rate: larger at the beginning, smaller at the end. This is more of an art than a science.

The **initial values** depend on the type of activation function and the loss function. We discuss here only the MSE. For the sigmoid activation function, it is important to initialize all weights and biases to small random values, so as to be in the linear range. For the ReLu activation function, it is important for at least one of the neuron inputs $\mathbf{z}$ to be positive. If they are all negative, the algorithm is stuck. The weights should be close to zero, and biases should be initialized to small positive values. For underfitted models, and whenever possible, initialize the parameters at the solution of a linearized version of the model.

Model averaging is the ML pendant of diversification in finance. If models have uncorrelated errors, it may make sense to average the results from each model. A typical application would be to average the same model across different initial values. As usual, there are counterexamples, as we shall see in Exercise 11.3: averaging results from different initial values may not always work.

We used a fixed number of iterations for the stochastic gradient algorithm. This is a simple **termination criterion**. Another choice is to stop at a critical point of the gradient of the full cost J (not at a critical point of the gradient of the reduced cost). In practice, the algorithm would stop when the norm of the gradient is sufficiently low. The problem is that this may never occur if the algorithm zigzags. Since an MLP may be stuck at a local minimum, it often does not make much sense for the algorithm to terminate at a critical point of the gradient. On the other hand, numerical errors may compound with too many iterations of the algorithm, as Figure 11.4 shows. **Early stopping** is a fairly simple termination criterion. The idea is to graph the cost function as a function of the number of iterations and stop the iterations as soon as the cost becomes much worse. Of course, the parameter $\hat{\theta}$ to be output is the minimizer of that curve. This is illustrated in Figure 11.4.

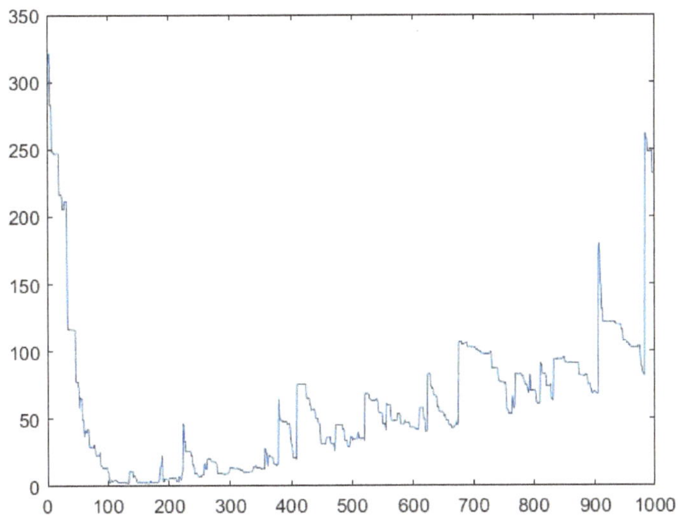

Figure 11.4: Cost of the MLP when applied to the single-neuron "network" described in Exercise 11.2. The cost is plotted as a function of the iteration number. The results get worse after 150 iterations, and early stopping would be optimal.

Exercises

11.1. Practitioners may be daunted by the complexity of neural networks. A popular alternative is to use functions that are linear in the parameters, but nonlinear in the data. A typical example is **radial basis function interpolation**, which we saw in Chapter 6. Consider the problem of learning the XOR function.

(a) Use Monte Carlo simulation to produce a sample of $T = 100$ datapoints $\{\mathbf{x}_t\}$ from the uniform distribution, where

$$\mathbf{x}_t = \begin{bmatrix} x_{1,t} \\ x_{2,t} \end{bmatrix}$$

Both $x_{1,t}$ and $x_{2,t}$ are real-valued and belong to the interval $[0, 1]$. The response function is

$$y_t = 1\left[x_{1,t} > \frac{1}{2}\right] 1\left[x_{2,t} > \frac{1}{2}\right]$$

(b) Suppose that the standard deviation σ is fixed. Write the normal equations for the OLS problem, where the fitted value is given by

$$\hat{y}_t = \sum_{t=1} \theta_t \exp(-||\mathbf{x} - \mathbf{x}_t||^2/2\sigma^2)$$

(c) For three different values of the hyperparameter σ, calculate the MSE of your estimator on your dataset. Choose for values $\sigma = 0.1$, $\sigma = 0.25$, and $\sigma = 1$.

11.2. **Early Stopping**
Apply the standard MLP with Relu activation, $L = 2$, on the following model:

$$y_t = Relu(x_t) + u_t$$

Use the following training dataset: the label x_t is generated at equal intervals from $[-5, 5]$, and the error in the dataset is not mean-zero; it satisfies

$$\hat{u_t} = \sigma\epsilon_t * Relu[x_t]$$

where ϵ_t is a standard normal variate. Fix $\sigma = 0.1$. Try to replicate Figure 11.4. We used the parameters $\sigma = 0.1, \texttt{rng}(1000)$, and initial value $(b, w) = (0.5, 1)$.

11.3. (Medium) **On a Pitfall of Model Averaging**
Consider the following model:

$$f^*(x) = |x|$$

You will apply the standard MLP with ReLu activation, $L = 3$, and two nodes in the hidden layer ($n_2 = 2$) to approximate it. Observe that the following two solutions are optimal:

$$f(x, \boldsymbol{\theta}^*) = Relu(W_{1,1}^{(2)*}x) + Relu(W_{1,2}^{(2)*}x)$$

with $W_{1,1}^{(2)*} = 1$ and $W_{1,2}^{(2)*} = -1$. Another solution is

$$f(x, \boldsymbol{\theta}^{**}) = Relu(W_{1,1}^{(2)**}x) + Relu(W_{1,2}^{(2)**}x)$$

with $W_{1,1}^{(2)**} = -1$ and $W_{1,2}^{(2)**} = 1$.

(a) Generate a dataset such that, with the initial condition $\mathbf{W}^{(2)}$ close to $\mathbf{W}^{(2)*}$, the algorithm converges to $\mathbf{W}^{(2)} = \mathbf{W}^{(2)*}$ while, with initial conditions $\mathbf{W}^{(2)}$ close to $\mathbf{W}^{(2)**}$, the algorithm converges to $\mathbf{W}^{(2)} = \mathbf{W}^{(2)**}$.

(b) Apply the standard MLP to the problem, and compare the value of the first estimator $\hat{\boldsymbol{\theta}}_1$ to $\boldsymbol{\theta}^*$ and of the second estimator $\hat{\boldsymbol{\theta}}_2$ to $\boldsymbol{\theta}^{**}$.

(c) Construct a new estimator:

$$\hat{\boldsymbol{\theta}}_3 = \frac{\hat{\boldsymbol{\theta}}_1 + \hat{\boldsymbol{\theta}}_2}{2}$$

Is it a good estimator?

11.4. Code an MLP to approximate the XOR function. Use the data from Exercise 11.1 and, if you did the latter, compare your results with radial basis function interpolation.

11.5. (Medium) Prove Lemma 11.2.
Hint: Show that

$$\frac{\partial z_k^{(l+1)}}{\partial z_j^{(l)}} = W_{k,j}^{(l+1)} \sigma'(z_j^{(l)})$$

11.6. Code an MLP to approximate the function $f^*(x) = \cos(x)$ for $0 \leq x \leq \pi/2$.
Hint: Approximate $\cos(x)$ by piecewise linear functions, which will be linear combinations of elementary functions of the form $Relu(x - \frac{m}{M}\frac{\pi}{2}) - Relu(x - \frac{m+1}{M}\frac{\pi}{2})$ for $0 \leq m \leq M - 1$. Choose $M = 5$, and graph manually $f(x, \boldsymbol{\theta}^*)$ as a function of x.

11.7. (Hard) Refer to Example 11.4, and show that a function f_{pos} satisfying (11.16) is given by

$$f_{pos}(t, \theta) = t - \sqrt{\left| \frac{dg_{pos}(t)}{dt} \right|}$$

where, for $b < 0$,

$$g_{pos}(t) = \ln \left[(b - (w - 1))^2 t + (2 - w)) \right] t$$

What is it for $b \geq 0$? Explain in words the meaning of what you are doing.

11.8. (Hard) **Representing a Complicated Function**
Refer to Example 11.4.

(a) Show that, for $b \geq 0$ and $s(t) = (b - (1 - w))^2(t + 1) + 2 - w$,

$$f_{neg}(t) = t - \sqrt{(b - (1 - w))^2(t + 1)/s(t) + \ln(s(t))}$$

(b) Show how to represent f_{neg} for $b \geq 0$ as a deep learning network by (i) reusing exercise 11.6, (ii) fixing parameters to be identical from one layer to the next, and (iii) representing the layer functions as functions of the two variables t and $s(t)$. Fixing parameters to be identical is used heavily in recurrent networks.

11.9. **Project: Building the Two-Valley Network**
Reuse the results from Exercises 11.7 and 11.8 and extend them to implement an MLP network that results in the two-valley network. Fix some parameters, and try to reproduce the table in Example 11.4.

Chapter 12

Long Short-Term Memory (LSTM) Networks

The feedforward network studied in the previous chapter handles only the case when the relation (11.1) maps a function of one sample point to one label. In **recurrent networks**, *previous* sample points $\mathbf{x}_{t-1}$ can have an impact on *current* labels $\mathbf{y}_t$. In other terms, the relation

$$\mathbf{y}_t = \mathbf{f}^*(\mathbf{x}_t) + \mathbf{u}_t \tag{12.1}$$

does not necessarily hold. The relation between input and output can be as complicated as

$$\mathbf{y}_t = \mathbf{f}^*(\mathbf{x}_t, \mathbf{x}_{t-1}, \dots, \mathbf{x}_0) + \mathbf{u}_t \tag{12.2}$$

One can see the problem immediately: because the dimension of the input space explodes, there can potentially be more parameters than sample points. Unless one imposes a restriction on the model (that is, restriction on the parameters), we will have overparameterization. **Recurrent networks** address this problem by forcing some parameters to be identical. LSTM networks are a type of recurrent networks that generalize to the nonlinear case the traditional (linear) time-series models from econometrics. We now turn to the latter.

12.1 ▪ Linear Time Series Models

As explained in the introduction, this book emphasizes cross-sectional models over time series models. Thus this chapter constitutes only a very brief introduction to this important class of models. We refer the reader to the books by Hamilton [19] or Tsay [25] for a complete description of linear models.

One of the main problems in a time series is **prediction**, or **forecasting**. In time series, the output is (a forecast) of the input at a later time, i.e., $\mathbf{y}_t \equiv \mathbf{x}_{t+1}$. The fundamental principle of estimation theory (Theorem 10.1) shows that the best predictor of $\mathbf{x}_{t+1}(\omega)$, in the least squares sense, is its conditional expectation $E_t[\mathbf{x}_{t+1}|\mathbf{x}_t, \mathbf{x}_{t-1}, \dots]$.

Autoregressive integrated moving average (**ARIMA**) models and their special case (**ARMA**) models are the foundations of time series. In this section, we will study only the scalar version of the time series (12.2), namely,

$$x_{t+1} = f^*(x_t) + u_{t+1}$$

where u_t is a collection of independent mean-zero normally distributed random variables with variance σ^2. The autoregressive of order 1 model, or **AR(1) model**, is the following:

$$f^*(x_t) = c + \phi x_t$$

Thus the parameter vector is

$$\boldsymbol{\theta}^{AR1} = \begin{bmatrix} c \\ \phi \end{bmatrix}$$

Putting everything together, the model is

$$x_{t+1}(\omega) = c + \phi x_t + u_{t+1}(\omega) \tag{12.3}$$

The AR(1) model is the simplest **mean-reverting** model. When applied to term structure modeling, it is called the (discrete-time version of the) **Vasicek** model of interest rates . When $c = 0$ and $\phi = 1$, we obtain the **random walk** model. Econometricians say that this model has a unit root.[34] By the assumed properties of the noise u_t, this random walk is a **martingale**.[35]

Clearly, this model is outside the general framework of feedforward neural networks—again, not because it is linear (linear is a particular case of nonlinear), but because two input data records (x_t and x_{t+1}) are used in the same relation. One can *solve* the AR(1) model recursively:

$$\begin{aligned} x_{t+1} &= c + \phi x_t + u_{t+1} \\ &= c + \phi(c + \phi x_{t-1} + u_t) + u_{t+1} \\ &= \cdots \\ &= c\sum_{i=0}^{\infty} \phi^i + \sum_{i=0}^{\infty} \phi^i u_{t+1-i} \end{aligned}$$

If $|\phi| < 1$, the series converges, and we have the **moving average representation**:

$$x_{t+1} = \mu + \sum_{i=0}^{\infty} \phi^i u_{t+1-i} \tag{12.4}$$

with

$$\mu = c\sum_{i=0}^{\infty} \phi^i = \frac{c}{1-\phi}$$

The model (12.4) is called the MA(∞) model. When $|\phi| < 1$ we say that the AR(1) process is **stationary**. If this is the case, and if the series has been running since $t = -\infty$, then the series is ergodic.

12.1.1 ▪ Statistics of the AR(1) Model

Comparing (12.4) with (12.3), we have

$$E[x_{t+1}] = \frac{c}{1-\phi}$$

The **autocovariance** function $\gamma(j) \equiv E[(x_t - \mu)(x_{t-j} - \mu)]$ in the AR(1) model does not depend on time t but on the difference in time between t and $t - j$. It can easily be calculated. We start with

$$(x_t - \mu)(x_{t-j} - \mu) = \sum_{i=0}^{\infty} \phi^i u_{t+1-i} \sum_{k=0}^{\infty} \phi^k u_{t+1-j-k}$$

[34]The presence of a unit root is detected by the Dickey–Fuller test.

[35]A martingale is a stochastic process for which today's value is equal to the conditional expectation of a stochastic process at any future value.

Since

$$E[u_j^2] = \sigma^2$$
$$E[u_j u_k] = 0 \quad j \neq k$$

the autocovariance function of the AR(1) model is

$$\begin{aligned}\gamma(j) &= \sigma^2 \sum_{i=0}^{\infty} \phi^i \phi^{i-j} \\ &= \frac{\sigma^2 \phi^{-j}}{1-\phi^2}\end{aligned}$$

Figure 12.1 shows different autocovariance functions. The first one has a positive ϕ, so that the autocovariance function is decreasing. The second one has a negative ϕ, so that the autocovariance function is oscillating. In both cases, since $|\phi| < 1$ the absolute variance of the autocovariance decreases with time, meaning that data far in the past has a small influence on the current value of the data. Most financial time series exhibit this pattern.

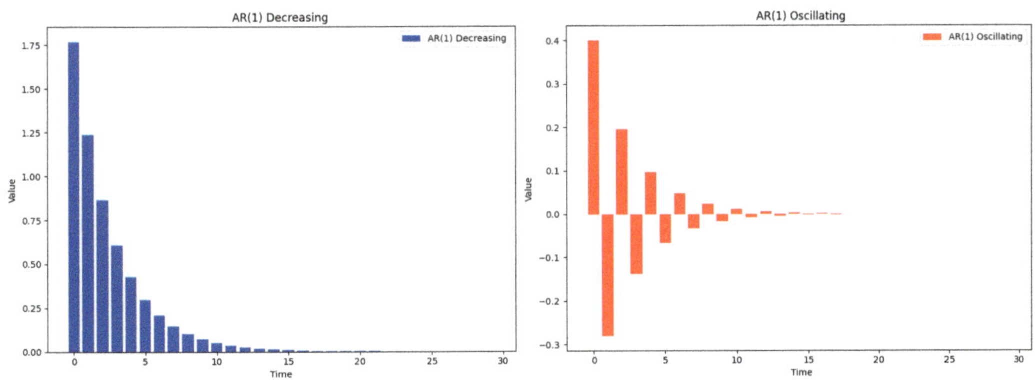

Figure 12.1: Autocovariance function of an AR(1) model.

A nice trick to find the joint probability density function $f_{x_1,\ldots,x_T}$ of $x_1, \ldots, x_T$ is to remember that $u_1, \ldots, u_T$ are IID normal random variables. Thus the joint density $f_{u_1,\ldots,u_T}$ of $(u_1, \ldots, u_T)$ is[36]

$$f_{u_1,\ldots,u_T}(u_1, \ldots, u_T) \sim \exp\left(-\frac{u_1^2}{2\sigma^2}\right) \exp\left(-\frac{u_2^2}{2\sigma^2}\right) \ldots \exp\left(-\frac{u_T^2}{2\sigma^2}\right)$$

Clearly

$$f_{x_1,\ldots,x_T}(x_1, \ldots, x_T) \sim f_{u_1,\ldots,u_T}(x_1 - c - \phi x_0, \ldots, x_T - c - \phi x_{T-1})$$

We always assume in this book that variables are known at time zero. In this case, we have

$$f_{x_1,\ldots,x_T}(x_1, \ldots, x_T) = \frac{1}{(2\pi\sigma^2)^{T/2}} \exp\left(-\sum_{t=0}^{T-1} \frac{(x_{t+1} - c - \phi x_t)^2}{2\sigma^2}\right)$$

[36]We chose the same notation for a random variable and the value that it takes. In this case, no confusion arises.

Defining the error $J(\boldsymbol{\theta})$ as minus the logarithm of this density function, we see that the method of moments coincides, in this case, with maximum likelihood estimation, as the following fact summarizes (see also Exercise 3.4).

Fact 12.1. *The maximum likelihood estimator of $\boldsymbol{\theta}$ minimizes the MSE.*

12.1.2 ▪ Implementation as an LSTM

The parameter vector $\boldsymbol{\theta}^{AR1}$ for the AR(1) process can be found analytically, as you will see in exercise 12.2. One can also use the following numerical approach:

- write the gradient of the MSE
- use the stochastic gradient algorithm to converge to a critical point of the gradient.

One then obtains the simplest LSTM network. You will see in Exercise 12.2 how to do it. In that problem, the LSTM network is a **preprocessor** for the MLP that does the nonlinear cross-sectional regression. In other words, the cross-sectional regression will be done on the residuals $\{\hat{u}_t\}$ of the AR(1) process and not on the time series $\{x_t\}$ themselves. If the cross-sectional model was linear, we could then invoke the Gauss–Markov theorem from Chapter 6 to prove that the resulting estimator is more efficient than if we had done a regression on the not transformed data $\{x_t\}$. This is the advantage of GLS over OLS: the error time series is more likely to be ergodic than the original series.

12.2 ▪ Nonlinear Time Series

Clearly, the linear relation (12.3) can be "nonlinearized" into

$$x_{t+1}(\omega) = f(x_t, \boldsymbol{\theta}) + u_{t+1}(\omega) \tag{12.5}$$

A typical neural network implementation is given in Figure 12.2. We refer the reader to Goodfellow, Bengio, and Courville [8] for the equations underlying this model. An analytically simpler alternative is to "force" the coefficients to be identical by regularization, as the following example shows.

Example 12.2. Take the simple case of the linear model $f(x_t(\omega), \boldsymbol{\theta}) = c + \phi x_t(\omega)$. Let us add a cross-sectional regression:

$$\breve{y}_{t+1}(\omega) = a + b x_t(\omega) + u_{t+1}(\omega)$$

Suppose we have $T = 4$ observations. We can define vectors:

$$\mathbf{x} = \begin{bmatrix} x_0 \\ x_1 \\ x_2 \\ x_3 \end{bmatrix}$$

and

$$\mathbf{y}' = \begin{bmatrix} \breve{y}_4 & \breve{y}_3 & \breve{y}_2 & \breve{y}_1 & x_3 & x_2 & x_1 & x_0 \end{bmatrix}$$

Clearly there is a linear relation:

$$\mathbf{y} = \mathbf{W}\mathbf{x} + \mathbf{d} + \text{noise}$$

where the $\mathbf{W}$ matrix not only is sparse but contains identical elements. For instance, $W_{1,5} = W_{2,6} = b$. Of course the neural network does not know that, but the designer does. One could then implement an MLP (with only one observation!) by adding regularization terms, i.e.,

$$\min_{\mathbf{W},\mathbf{d}} ||\mathbf{y} - \mathbf{W}\mathbf{x} - \mathbf{d}||^2 + \lambda_1(W_{1,5} - W_{2,6})^2 + \cdots \tag{12.6}$$

You will be asked in Exercise 12.4 to complete this example.

12.2.1 ▪ The LSTM as a Preprocessor

There are two main reasons to use an LSTM as a preprocessor in a regression. One, as we just saw, is to make the time series ergodic. The second one is to reduce the number of dimensions of the time series.

An example of the application of these ideas to finance is in the paper by Chen, Pelger, and Zhu [5] which we will describe fully in Chapter 26.

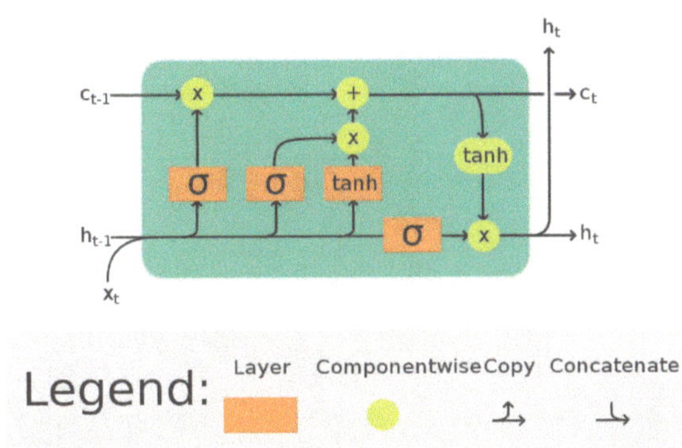

Figure 12.2: Typical LSTM network. Image by Guillaume Chevalier, distributed under a CC BY-SA 4.0 license https://creativecommons.org/licenses/by-sa/4.0/.

Exercises

12.1. (Hard) Prove Fact 12.1 (refer to Exercise 3.4).

12.2. (Medium) Find analytically the optimal parameters (c, ϕ) of the AR(1) model.

12.3. Given data $\{\mathbf{x}_t, y_t\}$, implement a neural network that minimizes

$$J(\boldsymbol{\theta}, \boldsymbol{\theta}^{AR1}) = \sum_{t=1}^{T} (y_t - f(\hat{\mathbf{u}}_t, \boldsymbol{\theta}))^2$$

where $f(\mathbf{x}_t, \boldsymbol{\theta})$ is the MLP that was discussed in the previous chapter and $\hat{\mathbf{u}}_t$ are the residuals of the realizations of the vector process

$$\mathbf{x}_t(\omega) = \begin{bmatrix} x_{t,1}(\omega) \\ x_{t,2}(\omega) \end{bmatrix}$$

Each scalar process $x_{t,i}(\omega)$ is an AR(1) process:

$$x_{t+1,i}(\omega) = c_i + \phi_i x_{t,i}(\omega) + u_{t,i}(\omega)$$

Assume that $\{u_{t,i}\}$ are independent normal random variables with mean zero and variance σ^2. We call $\boldsymbol{\theta}^{AR1}$ the parameters for this vector process, i.e.,

$$\boldsymbol{\theta}^{AR1} = \begin{bmatrix} c_1 \\ \phi_1 \\ c_2 \\ \phi_2 \end{bmatrix}$$

You should

(a) write the first-order conditions:

$$\frac{d}{d\boldsymbol{\theta}^{AR1}} J(\boldsymbol{\theta}, \boldsymbol{\theta}^{AR1}) = 0$$
$$\frac{d}{d\boldsymbol{\theta}} J(\boldsymbol{\theta}, \boldsymbol{\theta}^{AR1}) = 0$$

(b) implement this program in MATLAB by reusing the code for the MLP that was given in Exercise 2.2.

12.4. (Easy) Complete the formulation of the regularizer in the objective (12.6) in the example section.

Chapter 13

Generative Adversarial Networks (GANs)

Suppose that the objective is to minimize the cost function

$$\Omega(\boldsymbol{\theta}) = \sum_{T=1}^{T} h(\mathbf{x}_t; \mathbf{y}_t, \boldsymbol{\theta})^2$$

for some function h that depends on the data $(\mathbf{x}_t; \mathbf{y}_t)$ and the parameter vector $\boldsymbol{\theta}$. Suppose that we know a priori that $\boldsymbol{\theta}$ should satisfy a constraint of the form

$$\mathbf{g}(\mathbf{x}_t, \mathbf{y}_t, \boldsymbol{\theta}) \leq \mathbf{0}$$

for each $t = 1, \ldots, T$. A typical example in this book would be to boost the estimation of $\boldsymbol{\theta}$ by making the residuals across observations similar in size, like in the Cochrane–Orcutt algorithm described in Chapter 7. Define

$$\mathcal{S}(\boldsymbol{\theta}) = \sum_{t=1}^{T} \mathbf{g}(\mathbf{x}_t, \mathbf{y}_t, \boldsymbol{\theta})$$

One would then define a Lagrange multiplier $\boldsymbol{\lambda}$ and a Lagrangian:

$$L(\boldsymbol{\theta}, \boldsymbol{\lambda}) = \Omega(\boldsymbol{\theta}) + \boldsymbol{\lambda}'\mathcal{S}(\boldsymbol{\theta})$$

The saddle point theorem (Theorem 4.6) shows that this problem is equivalent to solving the dual problem

$$\min_{\boldsymbol{\theta}} \max_{\boldsymbol{\lambda} \geq \mathbf{0}} L(\boldsymbol{\theta}, \boldsymbol{\lambda})$$

How can we solve this problem? The Cochrane–Orcutt algorithm, as well as the primal-dual algorithm of convex programming, gives an answer: to alternate between solving the primal problem, i.e., minimize $L(\boldsymbol{\theta}, \boldsymbol{\lambda})$ over $\boldsymbol{\theta}$ for $\boldsymbol{\lambda}$ fixed, and maximizing $L(\boldsymbol{\theta}, \boldsymbol{\lambda})$ over $\boldsymbol{\lambda}$ for $\boldsymbol{\theta}$ fixed. Will this strategy converge to the solution? As always, it depends on the type of models considered.

The table in equation (4.25) provides the first-order conditions for this problem (for the case where $\boldsymbol{\theta}$ must be positive). In some cases the first-order conditions are sufficient for a saddle point to be a global optimizer, i.e., there is no *duality gap* (see Theorem 6.2.6 in Bazaraa and Shetty [16]). In other terms, an algorithm that converges to a solution of the first order (Kuhn–Tucker conditions) converges to an optimizer. We then have the following algorithm.

ALGORITHM 13.1. Gradient algorithm for GAN.

```
Set learning rate η, Nbiter
Set initial value for θ^0, λ^0
For iter =1 to NbIter
    Minimization
    1.Set k = 0
        Calculate L(θ^k, λ^0)
    2.Repeat
        Set θ^{k+1} = θ^k − η∇_{θ^k} L(θ^k, λ^0)
        k ← k + 1
    3.Until k =Nbiter
    4.Set θ^0 = θ^{Nbiter}
    Maximization
    5.Set k = 0
        Calculate L(θ^0, λ^k)
    6.Repeat
        Set λ^{k+1} = λ^k + η∇_{λ^k} L(θ^0, λ^k)
        k ← k + 1
    7.Until k =Nbiter
    8.Set λ^0 = λ^{Nbiter}
end
```

There are three loops in this algorithm. Of course the maximum number of iterations in each loop can be different. We chose the same value (`Nbiter`) for simplicity. This algorithm is typically viewed as a game between two adversaries, the minimizer and the maximizer, hence its name.

It turns out that this algorithm often helps regularize the parameters. The most obvious case is when we take $h(\mathbf{x}_t;\mathbf{y}_t,\boldsymbol{\theta})^2 = ||\boldsymbol{\theta}||^2$. Thus, so far we have three techniques for regularization:

- early stopping
- GAN
- ridge regression.

Exercises

13.1. (Easy) Put the Cochrane–Orcutt algorithm into the framework of a GAN. This is only an exercise in the translation of the notation from Chapter 7 into the notation of this chapter.

13.2. (Easy) Rewrite the GAN algorithm by substituting the stochastic gradient for the gradient algorithm.

13.3. Code in MATLAB (or any language of your choice) a version of a GAN where $\boldsymbol{\lambda}$ is a function of a parameter $\boldsymbol{\theta}_g$. To better distinguish between them (and to prepare for Chapter 26) we use the notation $\boldsymbol{\theta}_\omega$ for $\boldsymbol{\theta}$. We then write $\Omega = \Omega(\boldsymbol{\theta}_\omega)$ and $\mathcal{S} = \mathcal{S}(\boldsymbol{\theta}_g)$. The function Ω is the MLP from Example 11.1, while $\mathcal{S}$ is another MLP with $L = 3$ layers. Take $T = 1$ to keep your work manageable.

Chapter 14

Dynamic Programming and Reinforcement Learning

Dynamic programming is one of the techniques used to solve the **optimal control problem**. Optimal control has been used in many different areas, like rocket control, chemical process control, and optimal economic growth. We will see in Chapter 19 the application of this technique to solve the **strategic asset allocation** problem, both analytically and numerically.

In optimal control, the objective is to minimize a *cost functional* V of the *state variables* $x(t)$ of the system by acting on the *control variables* $u(t)$. The functional can also include the control variables. For instance, in rocket control, one wants to make sure a rocket lands on Mars ($x(T)$ is the position of the rocket at time T) while minimizing the amount of fuel used at each time t, namely, $u(t)$. The functional V will thus be, in this book,[37] a function of the state and control variables, measured at any time between now (time zero) and the *horizon* (denoted by T, possibly infinite). Three types of functionals have been extensively studied:

- in the *Lagrange* problem, the functional is a time integral of a function of the state and control variables
- in the *Mayer* problem, the functional is a function of the state variables at the horizon
- in the *Bolza* problem, the functional is the sum of the functionals of the previous problems.

It is not too difficult to translate one problem into another one by changing the definition of the state variables. We will focus on the Lagrange problem in this chapter and on the Mayer problem in Chapter 19.

Historically, the first technique used to solve the Lagrange problem was the calculus of variations. This method has been generalized, and now three techniques largely dominate the field:

- dynamic programming
- Pontryagin's maximum principle
- the martingale method, for the case of stochastic control.

We will not describe the maximum principle in this book and will focus on dynamic programming instead. The martingale method will be described in Chapter 19, since it is mostly used in strategic asset allocation.

We first describe deterministic dynamic programming and then move to stochastic dynamic programming.

[37]There are more general statements of the control problem. In a time-optimal problem, one would minimize the time T in which a rocket lands on Mars. We will not look at these formulations, which have fewer applications in finance.

14.1 • Deterministic Dynamic Programming

The **principle of optimality**, formulated by Richard Bellman [43], states that a trajectory of the state variable x is optimal only if each part of the trajectory is optimal. For instance, if the optimal path to go from A to C goes through B, then the part of that path that goes from B to C is also the optimal path for the reduced problem of going from B to C.

We will focus in this example on the Lagrange problem, where the problem is to minimize the objective V over all controls $u(t)$:

$$V(0, x(0); \text{controls}) = \sum_{t=0}^{T-1} I(t, x(t), u(t)) \tag{14.1}$$

for some function I called the reward function.

14.1.1 • Example: Shortest Path Problem

Figure 14.1 shows a directed graph, where the number on each edge represents the length of the edge linking two nodes in the plane. The problem is to find the shortest path between the start node and the end node.

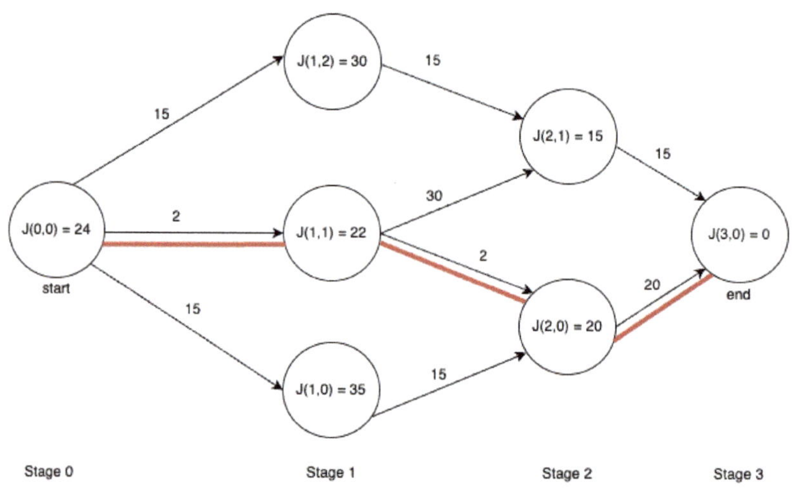

Figure 14.1: Backward induction in dynamic programming. Here $J(t, i)$ denotes the minimal distance from node (t, i) to the end node. The optimal path from the start node is in red.

The graph in this example has a particular **multistage** structure, whereby we can split the set of nodes into subsets, or stages, labeled by $t \in \{0, \ldots, T\}$, such that every directed edge from a node of stage t points to a node of stage $t + 1$. Each node can then receive a label (t, i), where t is the stage and, in our example, i corresponds to the "vertical" distance from the bottom. For instance, the bottom node in stage 1 is labeled $(1, 0)$, and the top node is labeled $(1, 2)$.

Suppose a traveler goes through the graph. The (vertical) position at stage t, namely, $x(t)$, is called the **state variable**. At each stage, she has the choice of going to one of several nodes in stage $t + 1$. This is the **control**. In this simple problem, we can label the control by giving possible destinations. For instance, the value that the control $u((1, 1))$ from node $(1, 1)$ takes belongs to the set $\{(2,0),(2,1)\}$. This example is an example of **feedback control**,[38] where the control at stage t, denoted by u, is a function of the state variable at stage t:

$$u(t) = u(t, x(t))$$

We associate a cost I for each edge. For instance, the cost of going from node $(1,1)$ to node $(2,1)$ is 30 kilometers, and we would write it as

$$I(1,1,(2,1)) = 30$$

The cost of going from node $(1,0)$ to node $(2,1)$ is infinite, and we would write it as

$$I(1,0,(2,1)) = \infty$$

We can now calculate the length of a path on a graph from node $(t, x(t))$ to the end node as

$$V(t, x(t);\text{path}) = \sum_{s=t}^{T-1} I(s, x(s), u(s, x(s)))$$

where we use the shorthand "path" to denote the values of the future controls. The problem is to calculate the shortest path from each node to the end node, or **value function**,

$$J(t,i) = \min_{\text{paths from } (t,i) \text{ to end}} V(t, i;\text{path})$$

and, in particular, $J(0,0)$, the minimal distance from the start node to the end node. Suppose the shortest path $J(t+1,k)$ has already been calculated for all nodes at stage $t+1$. The principle of optimality states that

$$J(t,i) = \min_k \{I(t,i,(t+1,k)) + J(t+1,k)\}$$

In other words, if the shortest path starting from node (t,i) goes through node $(t+1,k)$, then, from node $(t+1,k)$, it should take the optimal path to the end node. See Figure 14.1 for an illustration.

The **optimal trajectory** is denoted by $x^*(t)$.

ALGORITHM 14.1. Deterministic backward induction algorithm (Lagrange problem).

$J(T,0) = 0$
For $t = T-1$ to 0
  For all nodes (t,i) at stage t
    $J(t,i) = \min_k \{I(t,i,(t+1,k)) + J(t+1,k)\}$
    Denote by k^* the value of the minimizer
    Set $x^*(t+1) = k^*$
  End
End

Observe the key computational advantage: at each stage t, one optimizes the objective over the reduced path between t and $t+1$, and not over the full path ahead. One could also use forward induction, i.e., calculate all the shortest distances starting from the start node to any node at a subsequent stage. For deterministic systems, the direction (forward or backward) does not matter.

[38]Note that not every problem has a feedback control. For instance, it may take the same amount of time to hike from the start node to node (t,i), using path A or path B, but path A is much more tiring than path B because path A goes through a hot climate. As a result, it will take more time to go to any subsequent node using path A than path B.

14.2 ▪ Stochastic Dynamic Programming

We now denote by $x^{\mathbf{u}}(t,\omega)$ the **state variable** at time t along scenario ω when subject to the control $\mathbf{u} = (u(0),\dots,u(T,\omega))$. Clearly the state variable at time t can be influenced only by the control at times before t.

Definition 14.1. *We write $\mathcal{U}$ for the set of values that $u(t)$ can take and for $\mathcal{X}$ for the set of values that $x(t)$ can take.*

In full generality, the number of outcomes at time t should thus be of order (card($\mathcal{X}$)$*$ card($\mathcal{U}$))^t: previous values of the state and of the control. In order to prevent this explosion in the number of outcomes, we restrict our approach to **Markov decision processes**, i.e., processes where the future values $x^{\mathbf{u}}(t+1,\omega)$ depend only on the current value of the state $x^{\mathbf{u}}(t,\omega)$, the current value of the control $u(t,\omega)$, and some noise, denoted by $z(t+1,\omega)$, which is unknown at time t but known at time $t+1$. Moreover, the collection of random variables z_t is IID. Moreover, the control is in **feedback form**, i.e.,

$$u(t,\omega) = u(t, x^{\mathbf{u}}(t,\omega))$$

The resulting state equation is then

$$x^{\mathbf{u}}(t+1,\omega) = f(t, x^{\mathbf{u}}(t,\omega), u(t, x^{\mathbf{u}}(t,\omega)), z(t+1,\omega)) \tag{14.2}$$

The problem is, given a horizon T, to minimize the cost functional $V^{\mathbf{u}}(0,x(0))$, where, in the Lagrange problem,

$$V^{\mathbf{u}}(t,x) = E\left[\sum_{s=t}^{T-1} I(s, x^{\mathbf{u}}(s), u(s))] | x^{\mathbf{u}}(t,\omega) = x\right] \tag{14.3}$$

Clearly, $V^{\mathbf{u}}(t,x)$ depends only on the current control $u(t)$ and the future (random) controls $u(t+1,\omega),\dots,u(T,\omega)$. We define the **value function** as

$$J(t,x) = \min_{\mathbf{u}} V^{\mathbf{u}}(t,x)$$

Example 14.2. The cost $I(t)$ could be (minus) the discounted value of the dividends of the portfolio at time t and $V^{\mathbf{u}}(t,x)$ the value of a portfolio following future policy $\mathbf{u}$.

When the minimizer is unique, we have the following definition.

Definition 14.3. *The optimal control $\mathbf{u}^*$ is the function of time and state that minimizes the cost $V^{\mathbf{u}}(0,x(0))$ over $\mathbf{u}$.*

At this point, it seems that at time t one has to evaluate a number of control variables equal to card($\mathcal{X}$)$^{T-1-t}$, each of them having card($\mathcal{U}$) values, resulting in a total of (card($\mathcal{U}$)$*$ card($\mathcal{X}$))$^{T-t-1}$ different trajectories to investigate. Fortunately, the principle of optimality can be applied.

Theorem 14.4. *Suppose that (14.2) holds; then the number of control variables u to be calculated at time t is equal to card$\{\mathcal{X}\}$.*

Proof. Since $I(t,x,u(t))$ is known at time t, we can take it out of the conditional expectation (14.3) and write

$$V^{\mathbf{u}}(t,x) = I(t,x,u(t,x)) + E[V^{\mathbf{u}}(t+1, x^{\mathbf{u}}(t+1)) | x^{\mathbf{u}}(t) = x] \tag{14.4}$$

The principle of optimality tells us that every part of the trajectory is optimal; thus

$$J(t,x) = \min_{u(t,x)} \left[I(t,x,u(t,x)) + E[\underbrace{\min_{u(t+1),\dots,u(T-1)} V^{\mathrm{u}}(t+1,x^{\mathrm{u}}(t+1))}_{J(t+1,x^{\mathrm{u}}(t+1))} | x^{\mathrm{u}}(t) = x] \right] \tag{14.5}$$

Writing u instead of $u(t,x)$ for simplicity, we arrive at the **Hamilton–Jacobi–Bellman (HJB) equation**:

$$J(t,x) = \min_{u} \left[I(t,x,u) + E[J(t+1,x^{\mathrm{u}}(t+1)) | x^{\mathrm{u}}(t) = x] \right] \tag{14.6}$$

■

The HJB equation (14.6) can be directly translated into an algorithm, where we calculate the value function backward. Observe that, unlike the deterministic case, we cannot calculate it forward: (14.4) would not work.

ALGORITHM 14.2. Stochastic backward induction algorithm.

```
J(T,x) = 0 for all values of x
For t = T − 1 to 0
    For all values of x ∈ 𝒳
        For each u ∈ 𝒰
            Calculate E[J(t+1, x^u(t+1))|x^u(t) = x]
        J(t,x) = min_u [I(t,x,u) + E[J(t+1, x^u(t+1))|x^u(t) = x]]
        Store the value of the minimizer u*(t,x) = u
    End
End
```

The main additional complexity in the stochastic problem, compared to the deterministic one, is that the algorithm needs to calculate conditional expectations at time t of the value function at time $t+1$, for each value of the state variable $x(t,\omega)$ and of the control u, as shown in Figure 14.2. The number of such computations is $T{*}\mathrm{card}(\mathcal{U}){*}\mathrm{card}(\mathcal{X})$. An alternative is to postulate a functional form of $u^*(t,x) \equiv g(x;\boldsymbol{\theta}^*(t))$ and to "learn" the optimal parameters $\boldsymbol{\theta}^*(t) \in \mathcal{T}$, the space of possible values of the parameters.

The problem is made easier if the optimal parameter vector $\boldsymbol{\theta}^*$ does not depend on time. We thus replace the notation x^{u} by the notation $x^{\boldsymbol{\theta}}$. In this case, a least squares Monte Carlo algorithm can be more efficient than a standard calculation of the conditional expectation, as Figure 14.3 shows.

14.2.1 ▪ Least Squares Monte Carlo (LSMC)

In terms of complexity, it can be quite hard to calculate conditional expectations for all possible values of $\mathbf{x}(t,\omega)$ and all values of the parameters when the dimension of $\mathbf{x}$ is large. The general principle of estimation (Theorem 10.1) tells us that if $x^{\boldsymbol{\theta}}$ is a Markov process, the conditional expectation $E_t[J^{\boldsymbol{\theta}}(t+1,x^{\boldsymbol{\theta}}(t+1))]$ solves the problem

$$\min_{\text{function}} E[(J^{\boldsymbol{\theta}}(t+1,x^{\boldsymbol{\theta}}(t+1)) - \text{function}(x^{\boldsymbol{\theta}}(t),t))^2]$$

For instance, we take for our set of functions an expansion in the polynomial functions

$$\text{function}(x^{\boldsymbol{\theta}}(t),t) = b_0(t) + b_1(t)x^{\boldsymbol{\theta}}(t) + b_2(t)(x^{\boldsymbol{\theta}}(t))^2 + \cdots$$

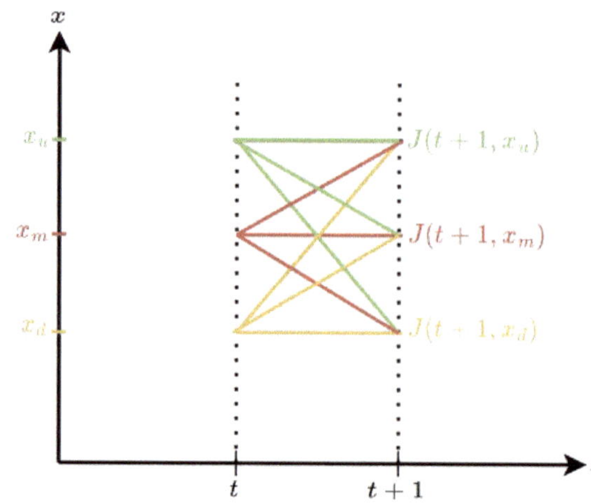

Figure 14.2: Fix a control u. The state variable $x(t)$ is a Markov chain that can take three possible values: x_u, x_m, x_d. The arrows represent all the transition probabilities. For instance, the top arrow represents $P(x(t+1) = x_u | x(t,\omega) = x_u)$. In the standard algorithm, we calculate the conditional expectation of the value function for all values of $x(t,\omega)$. For instance, $E[J(t+1, x(t+1)) | x(t,\omega) = x_u] = \sum_y P(x(t+1) = y | x(t,\omega) = x_u)\ J(t+1, y)$.

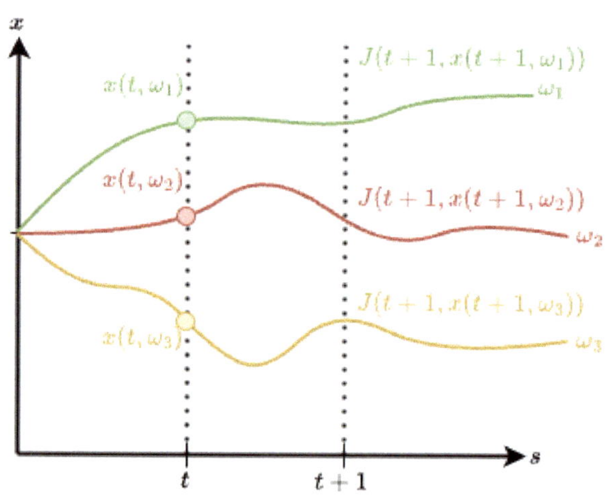

Figure 14.3: There are three scenarios, ω_1, ω_2, and ω_3. In the LSMC method, the conditional expectation of $J(t+1, x(t+1))$ is approximated by one regression of $J(t+1, x(t+1,\omega))$ on $x(t,\omega)$.

where $b_0(t), b_1(t), b_2(t), \ldots$ are deterministic, and we replace the population problem by the sample problem

$$\min_{b_0(t) b_1(t), \ldots} \sum_{\omega=1}^{\Omega} (J^{\boldsymbol{\theta}}(t+1, x^{\boldsymbol{\theta}}(t+1,\omega)) - (b_0(t) + b_1(t) x^{\boldsymbol{\theta}}(t,\omega) + \cdots)^2$$

This is a linear regression. We denote the resulting approximation by

$$\hat{E}_t[J^{\boldsymbol{\theta}}(t+1,x^{\boldsymbol{\theta}}(t+1))](\omega) \equiv b_0^*(t) + b_1^*(t)x^{\boldsymbol{\theta}}(t,\omega) + \cdots$$

This calculation is illustrated in Figure 14.3, which you should compare with Figure 14.2.

We then obtain the following algorithm.

ALGORITHM 14.3. Least squares Monte Carlo backward induction algorithm.

```
Initialize x(0)
For t = 0 to T − 1
    For all scenarios ω
        Simulate z(t+1,ω)
        For all possible values of θ
            Calculate u(t,x(t,ω)) = g(x(t,ω);θ)
            x^θ(t+1,ω) = f(t,x^θ(t,ω),u(t,x^θ(t,ω)),z(t+1,ω))
        End
    End
End
For all possible values of θ
    J^θ(T,ω) = 0 for all ω
    For t = T − 1 to 0
        Regress J^θ(t+1,x^θ(t+1,ω)) on powers of x^θ(t,ω)
        J^θ(t,x^θ(t,ω)) = I(t,x^θ(t,ω)) + Ê_t[J^θ(t+1,x^θ(t+1))](ω)
    End
End
Set θ* = arg min_θ J^θ(0,x(0))
```

This algorithm has a forward pass and a backward pass. The latter is more computationally intensive. The number of regressions is $T{*}\text{card}(\mathcal{T})$. We can thus compare the complexity of the two algorithms we have covered. See Table 14.1.

	Numerically intensive calculations	Number of such calculations
Backward induction	Cond. expectation	$T{*}\text{card}(\mathcal{T}){*}\text{card}(\mathcal{X})$
LSMC backward induction	Regressions	$T{*}\text{card}(\mathcal{T})$

Table 14.1: Complexity of two stochastic control algorithms.

Clearly, one could organize calculations differently. Rather than doing an exhaustive evaluation of the parameter space $\mathcal{T}$ like above, one could use a gradient algorithm.

14.2.2 ▪ Separability

There is often uncertainty on the parameters of the models. For instance, we will see in Chapter 24 that the betas can be modeled as a partially observable process and estimated by the *Kalman filter*. However, the betas are parameters of a portfolio optimization problem. Can we decompose the problem into

- estimating the betas?
- optimizing the portfolio?

Separation theorems answer in the affirmative. In some cases (such as linear models), the optimal solution to a problem can be obtained by taking for (a component of) the state process the best estimate of a partially observable process. We refer the reader to Oksendal [44] for further discussion.

14.2.3 ▪ Analytical Solutions

We will see in Chapter 19 how to solve the control problem analytically for the Merton [45] problem by dynamic programming. We will also show how to solve it by the martingale method, which is sometimes simpler than dynamic programming.

14.3 ▪ Reinforcement Learning

We now return to machine learning. We thus do not assume a parametric model of the state; i.e., we do not know f in (14.2). As a result, we cannot calculate the conditional expectation (14.3). We only observe values taken by the state variable $x^{\mathbf{u}}$. As in dynamic programming, the state variable depends on the **action** or control u taken. Nature is kind enough to provide us with the value $x^{\mathbf{u}}(t+1)$ for a large quantity of values of the control. We assume that the following variables are independent of time:

- the cost $I = I(x(t), u(t))$
- the value function $J = J(x(t))$.

Methods tend to be of the value iteration type and not of the backward induction type. Let us return to the HJB equation with stationary value function

$$J(x) = \min_u \{I(x,u) + e^{-r_f} E[J(x^{\mathbf{u}}(t+1))|x^{\mathbf{u}}(t) = x]\} \tag{14.7}$$

Note that we added a discount factor in the equation. If $r_f = 0$, this is the same equation as the stationary version of (14.6). Several algorithms in reinforcement learning are inspired by the traditional dynamic programming method of **value iteration**. In that method, one tries recursively to build a function $J : \mathbb{R} \to \mathbb{R}$, hoping to converge to a fixed point.

ALGORITHM 14.4. Value iteration algorithm.

1. `Set` $k = 0$, $J^0 = 0$
2. `Repeat`
   `For all` x
   $J^{k+1}(x) = \min_u \{I(x,u) + e^{-r_f} E[J^k(x^{\mathbf{u}}(t+1))|x^{\mathbf{u}}(t) = x]\}$
   $k = k+1$
   `End`
3. `Until converge`

Note that a "safer" update (in terms of convergence) is to replace J^k by a convex combination of J^k and J^{k+1}:

$$J^{k+1}(x) = (1-\eta)J^k(x) + \eta \min_u \{I(x,u) + e^{-r_f} E[J^k(x^{\mathbf{u}}(t+1)|x^{\mathbf{u}}(t) = x]\} \tag{14.8}$$

where the *learning rate* η is between zero and one.

14.3.1 ▪ Temporal Difference Method

In temporal difference (TD) methods, one fixes the control, and the conditional expectation (14.3) is replaced by the current value function of the next state $J(x^u(t+1))$. The equivalent of (14.8) is then

$$J^{k+1}(x) = (1-\eta)J^k(x) + \eta \min_u \{I(x,u) + e^{-r_f} J^k(x^u(t+1))\} \tag{14.9}$$

Of course, when only one value of u is available in the data, one dispenses with the minimization. After several iterations, time averaging will take place. If the system is ergodic, time averages will converge to expectations.

14.3.2 ▪ Q-Learning

Compared to TD, we reintroduce in *Q-learning* optimization over the future control. For convenience, we split the control into its value now and its value in the future.

Notation: Fix t. We define the current control a and the future control π by

$$a = u(t) \qquad \pi = \begin{bmatrix} u(t+1) \\ \vdots \\ u(T) \end{bmatrix}$$

Rather than working with the value function, one works with the *action-value function* $Q^\pi(x,a)$, which we introduce now.

Definition 14.5. *The action-value function for a policy* $\mathbf{u} = [\ a \quad \pi\ ]$ *is defined as*

$$Q^\pi(x,a) = E\left[I(x,a) + \sum_{i=1}^{T-t-1} e^{-r_f i} I(x^{\mathbf{u}}(t+i), u(t+i)) | x^{\mathbf{u}}(t) = x \right]$$

Corresponding to an optimal future policy π^*, there is an *optimal action-value function* Q^* which satisfies the HJB equation:

$$Q^*(x,a) = E[I(x,a) + e^{-r_f} \min_{a'} Q^*(x^{\mathbf{u}}(t+1), a') | x^{\mathbf{u}}(t) = x] \tag{14.10}$$

In Q-learning (like in the TD method), conditional expectations in (14.10) can be replaced by a sample value. The equation corresponding to the TD update (14.9) is

$$Q^{k+1}(x,a) = (1-\eta)Q^k(x,a) + \eta(I(x,u) + e^{-r_f} \min_{a'} Q^k(x^{\mathbf{u}}(t+1), a')) \tag{14.11}$$

Various schemes can be imagined, such as, rather than always taking the same value $x = x(t)$, taking x observed at different times.

14.4 ▪ To Probe Further

The material on deterministic dynamic programming and control is standard. A good introductory textbook is Intriligator [17]. Stochastic control is a difficult area of applied mathematics. It is usually studied in continuous time. For readers familiar with stochastic calculus, we recommend the book by Pham [46]. The least squares Monte Carlo algorithm was introduced by Longstaff and Schwartz [47] and Tsitsiklis and Van Roy [48] in the context of American option pricing. An interesting comparison between these algorithms is described in Glasserman [49].

There is a large number of methods that we did not have the space to cover. Among others, for Markov decision processes, is the policy iteration method. Among machine learning methods is the G-learning method, where the policy is stochastic: it assumes that the decision-maker makes random decisions. We refer the reader to the book by Dixon, Halperin, and Bilokon [50].

It remains to be seen whether reinforcement learning will be widely adopted in asset management. The industry tends to be conservative and has been slow to accept well-established methods, like dynamic programming. Will it move to more esoteric methods, whose results are less clear to interpret than results from dynamic programming?

Exercises

14.1. (Easy) Consider the Bolza problem

$$\min_u \sum_{t=0}^{T-1} c(t)x(t) + B(x(T))$$
$$x(t+1) = f(x(t), u(t))$$

Transform it into a Mayer problem; i.e., define a function U, a process $\mathbf{y}$, and a vector-valued function $\mathbf{g}$ such that

$$\min_u U(\mathbf{y}(T))$$
$$\mathbf{y}(t+1) = \mathbf{g}(\mathbf{y}(t), u(t))$$

returns the same result.

14.2. (Medium) Calculate the complexity of the LSMC backward induction algorithm when the parameter vector $\boldsymbol{\theta}$ depends on time, i.e.,

$$u(t,\omega) = g(x(t,\omega); \boldsymbol{\theta}(t))$$

14.3. (Hard) Solve the problem

$$\min \sum_{t=1}^{T} w(t)u(t)^2$$
$$u(t) \geq 0 \quad t = 1, \ldots, T$$
$$\sum_{t=1}^{T} u(t) = c$$

by dynamic programming.

14.4. (Hard) Solve the nonlinear programming problem

$$\min x_1 x_2 \ldots x_I$$
$$x_i \geq 0$$
$$\sum_{i=1}^{I} x_i = c$$

by dynamic programming.

14.5. Implement Algorithm 14.1, and apply it on the data of Example 14.1.1.

14.6. Code the least square Monte Carlo backward induction algorithm. Assume that $\{z(t)\}$ is a collection of independent standard normal random variables and that

$$x(t+1) = c + \phi x(t) * u(t) + z(t+1)$$
$$x(0) = 0$$

with $c = 1$ and $\phi = 0.9$. The reward function is

$$I(t, x(t), u(t)) = x(t)^2 + u(t)^2$$

Take $\Omega = 1000$ scenarios, and use a quadratic approximation of the value function.

Part II

Asset Management and Pricing

Chapter 15

Asset Management Basics

We first look at who invests, then what they invest in, and then how. The last section about the future of the field reflects only the opinion of the authors and can be skipped, as it is based more on common sense than science. And as Descartes said, "*le bon sens est la chose du monde la mieux partagée*."[39]

15.1 ▪ Asset Owners and Intermediaries

In most cases, the ultimate asset owners/investors are individuals.[40] Unlike a casino, where gambling rules are simple, investing can be quite complicated. For that reason, individuals most often delegate some part of the investment decisions to *financial intermediaries*, whether they are banks, mutual funds, hedge funds, pension funds, sovereign funds, or insurance companies. We call all of the latter **asset managers** or **portfolio managers**. The term **wealth management** currently refers to the case where the financial advisor works jointly with the individual investor, whether the individual is wealthy or not.

Asset owners mostly differ in their income, level of risk aversion, horizon, and types of liabilities they face. In the US, most salaried workers earn more and more as they get closer to retirement age, but then their retirement benefits (pension or 401(k)) tend to be lower than their salary. Liabilities also differ across a lifetime. In the US, many services, such as education and health, are not completely provided for by the state and are significant, uncertain liabilities. For that reason, a common investment strategy recommended by financial advisors is to gradually decrease the stock holdings in an investment portfolio: the risk of being destitute in old age and unable to cover medical expenses is just too large. In terms of modeling, the **horizon** of the investment decreases with age. **Insurance** can mitigate risk. It works along two different principles. One is the principle of mutuality: a large insurance pools individual risks together. If one insured person suffers a large covered loss, this loss is likely to be offset by the premium collected on other insured persons, who are unlikely to have the same accident at the same time. This is the principle of **diversification**. If risk cannot be diversified away, financial institutions can trade the residual risk.

For an investor, there are diverse levels of risk, and thus of expected return, in placing her money with a different **intermediary**. Banks, up to a certain amount, are probably the safest. In many countries, like the US, the government insures deposits up to a certain amount. This is

[39]Meaning that every human is endowed with the same amount of common sense.

[40]Here are some exceptions: (a) traditional US pension plans, where individual participants are creditors of the plan; (b) endowments; (c) sovereign wealth funds, where individual voters do not own the assets.

one of the reasons why the expected return on a deposit is very low. Mutual funds tend to offer simpler, less risky investment strategies than hedge funds. For that reason, hedge funds are open only to wealthy investors. Large insurance companies, pension funds, and sovereign funds do not put their large assets in a bank. They tend to invest themselves.

What is the best way to ensure that financial intermediaries act in the interest of the investor? One is regulation. Contract theory can also help. There is now a substantial literature on delegated portfolio management, which is part of the principal-agent problem in economics. For state-of-the-art measures of how active a portfolio manager is, see [51].

These stylized features of investor/intermediary financial behavior in the US are considered mainstream knowledge. Most other findings are controversial or country-specific, and we refer the reader to more specialized literature.

15.2 ▪ The Market: Investment Vehicles

In this book, we cover three different types of investments: stocks, bonds, and options. As the main focus of this book is not investment vehicles, we limit ourselves to describing them as instruments providing payoffs at different times and do not go much into investment frictions. The payoff of these instruments can be random (for **risky** securities: dividends for stocks, option payoff) or not random (for **fixed income** securities: risk-free bonds). There are clearly many other types of investment: commodities, real estate, and foreign investments. Yet their prices can be described with the same features as stocks: expected returns, volatilities, and correlations.

The financial industry keeps inventing new instruments. They can result in a greater social good (most options) or not (private cryptocurrencies, particularly in countries that have a stable currency and payment system). The most useful instruments, in our opinion, remedy a lack of **liquidity** in the market. With options, investors can better trade a systematic risk, i.e., a risk that cannot be diversified away, than by using the spot market. Also, options are an easy avenue for **short-selling**: whereas it is possible to short a security (i.e., sell a security one does not own), it is often indeed quite costly (Cochrane [21], Chen and Zhou [52, Chapter 1]).

The 2010 decade saw an extreme peak of **high-frequency trading**. Taking advantage of market illiquidity was quite a large business in the US. Indeed, in 2011, 2% of the 20,000 trading firms in the US were thought to be high-frequency traders and were responsible for 75% of all of US equity trading (Kolm and Maclin [53]). Since then, practitioners have been using more refined models to take trading costs into account, as we will see in Chapter 19. This extreme form of trading is controversial. It can be contested on moral and practical grounds: if the information is gradually and efficiently processed by the market, what is the need to trade every millisecond?

In this book, we do not offer a special treatment of credit risk. **Credit risk**, i.e., the risk of a corporation defaulting on its bonds, can be handled mathematically in the same way as market risk. A risky bond is just a security with a random payoff that trades at a discount compared to an equivalent risk-free bond. For special investments, like over-the-counter, uncollateralized derivatives, one must be aware of **counterparty risk**, i.e., the risk that a financial intermediary will default.

Other sources of friction are of macroeconomic nature: they include taxes, foreign exchange, and inflation. Some investments, like real estate, are more regulated than others.

15.3 ▪ Current Practices and Strategies

The asset management industry seems to prefer portfolios with a smaller number of different assets to larger ones. Whereas academics may use the full 32,000 stocks contained in the CRSP database, “Institutions hold relatively few stocks. The median institution held 67 stocks in the period from 2015 to 2017" (Brunnermeier et al. [54]). Hedge funds such as Renaissance, Two Sigma, and D.E. Shaw make thousands of trades and manage tens of billions of dollars (Goldstein, Spatt, and Ye [55]), which is just a fraction of what the largest wealth manager, UBS, holds.

Yet every asset manager offers a different strategy. We saw three general types of strategy in the introduction: technical analysis (used more by speculators than by large asset managers), fundamental analysis, and model-driven investing, of which pricing by arbitrage is an important component.

Some strategies exist because it would be too difficult for the individual investor to implement them efficiently. **Index tracking** belongs to that type of strategy. An investor who wants to replicate the S&P 500 index will find it difficult to invest in 500 stocks at the same time. Other strategies involve financial options; they exploit the mathematical difficulty of hedging them perfectly[41] (see Kakushadze and Serur [56]). Many strategies would exploit the fact that some assets offer a risk premium, represented by a **risk factor**. For instance, inflation is a risk factor. The inflation risk premium exists because most investors (like salaried workers) are exposed to it, and can hardly hedge it, while few others (like T-bill traders) are not. The definition of "bad times" is thus different for salaried workers (for them it is high inflation times) than for traders. As Ang [6] states, "Investors generally dislike the risk of losses during bad times," thus justifying risk premia.

From a theoretical standpoint, linear factor models can be derived directly from the **capital asset pricing model (CAPM)** and its generalization, the arbitrage pricing theory. Here there is a traditional division of labor and of goals between academics and practitioners. Many academics have been intent on proving the **market efficiency hypothesis**. Under that hypothesis the alpha of each security should be zero. The more factors one includes, the more likely alpha, i.e., riskless superior expected return, is to be zero. The opposite view is that markets are not efficient and that many securities are **mispriced**. A model with few factors will thus naturally result in several securities having a nonzero alpha. The rule is then to buy securities with a positive alpha and sell securities with a negative alpha. Identifying the correct factors is thus key; it is often done with data mining. Fama [57] is one of the most important proponents of these factor models, but he himself criticized their overuse, calling them a "fishing license." At the time of writing this book, the most popular factors in the US were market return, firm size, book-to-market ratio, momentum, and volatility. This may well change over the next years.

The only factor that is hard to dispense with, at least in theory, is the market return, which was historically the first factor. Going back 10 years before the CAPM was established, Markowitz [58] pioneered **mean-variance analysis**, which was the start of modern portfolio theory. Portfolios constructed using mean-variance analysis generally either have poor performance and/or change considerably over time, thus generating large transaction costs. The main reason for poor performance comes from estimation problems: the mean and variance of security returns are hard to estimate, and they can change fairly rapidly. In part for that reason, the industry moved to Bayesian methods for statistical estimation. One of them, the **Black–Litterman** method, not only provides more robust estimates but enables the investor to specify subjective views on asset returns (for instance, Microsoft will do 5% better than other stocks) in a consistent, risk-adjusted manner.

We move now from static (classical mean-variance) to **dynamic portfolio choice**. In the latter, a common goal is to maximize the expected utility of wealth over a long time horizon while rebalancing the portfolio at fixed dates between now and the horizon. We can distinguish

[41]This is a risky zero-sum game. If you are tempted by it, consider the fact that your counterparty will be Goldman Sachs or another Wall Street firm, which have collectively hundreds of years of experience in the domain. The odds are worse than in Las Vegas.

between two types of approaches: tactical and strategic asset allocation.[42] In **tactical asset allocation**, the optimal proportion of equities in the portfolio (called sometimes the *short-term weight*), measured in total dollars, is equal to the Sharpe ratio of equities divided by the product volatility of equities times degree of risk aversion. If the mean and variance of equities are assumed ex ante not to change over time, the short-term weight coincides with the *long-term weight*, that is, the proportion that static mean-variance analysis would have determined over a long horizon. Even if the short-term weight turns out to be constant ex post, a tactical investor must still rebalance his portfolio periodically, that is, sell stocks if stocks go up, and buy stocks if stocks go down, to keep a constant weight. It is thus a *contrarian* investment. In **strategic asset allocation** the investor does not assume ex ante that the parameters of the model are constant. One can incorporate stochastic volatility and stochastic interest rates. The resulting allocation tends to be even more contrarian that the short-term weight in tactical asset allocation and harder to justify to a CIO (Chief Investment Officer) if he/she happens not to be contrarian.

The **risk-neutral methodology** is used extensively to price financial derivatives. The skills required to price derivatives are different (often more mathematical) than the skills required for portfolio management in industry: few time series and many time steps in the former, and multiple securities but single horizon in the latter. Yet they share the same fundamental financial principle: no riskless arbitrage. When large cross-sections are involved, academics prefer to call this method the stochastic discount factor (SDF) methodology. Academics apply it to asset pricing pricing. This book describes several attempts to extend it to portfolio management. The **SDF methodology** presented in Chapter 23 is more general than the CAPM because it can be tailored to stronger assumptions: absence of arbitrage, which is generally held to be true in some markets, like derivative markets. Yet, like the CAPM, the SDF methodology also delivers a fairly simple portfolio construction methodology that identifies the portfolio with the highest Sharpe ratio.

Banks and insurance companies are as much financial assets managers as financial liabilities managers. For that reason, these firms sometimes integrate these techniques into (financial) **asset and liability management**, meaning that they jointly optimize their assets and their liabilities.[43] Traditionally, however, asset management and liability management divisions were separate units within banks and insurance companies. For that reason, **risk management** divisions were developed, the function of which is, once the investment and funding decisions have been made, to manage the residual risk separately. The main technique in financial risk management is **hedging**, meaning buying/selling a portfolio of securities (most often derivatives), the value of which will go in the opposite direction than the bank's portfolio. The question of how much a firm should hedge is outside the scope of this book. For nonfinancial firms, this is a classical corporate finance question.

The biggest hurdle facing the deployment of modern machine learning techniques is **lack of interpretability**. "Suppose somebody offers you two investment plans. Plan A offers a 10-year success rate of 70% and is a strategy based on fundamentals that can be readily understood. Plan B is marketed as a strategy that has a 10-year success rate of 80% but is a black box[...] Most rational investors would likely prefer Plan A to Plan B" (Aaron Low, in the Foreword of [52]). For methods to alleviate this lack of interpretability, see Chen and Zhou [52, Chapter 1].

15.4 ▪ Lessons from the Past: Useful for the Future?

History tends to repeat itself. Investors forget. Real estate in the US is a typical example of a market with irrational booms and crashes. Here is another one, which investors sometimes forget.

[42]To keep things simple in this introductory chapter, we focus only on the optimal ratio of equities in a portfolio.

[43]In the middle of the Euro crisis, when Euro-labeled money needed a safe haven, the Zurich Kantonal Bank in Switzerland refused to take more deposits. This bank acknowledged it did not know how to invest these extra deposits profitably. Such prudence did not encumber Silicon Valley Bank.

Banks and corporate treasuries of financial institutions have a natural propensity to use short-term money (which is cheaper than long-term bonds) to fund long-term investments. Without proper asset/liability management, whose goal is to match maturities, this leads to repeated crashes: Orange County in 1994, and Silicon Valley Bank in 2023. One would have thought that proper asset/liability management could have been learned in 1994 and applied all the way to 2023. Yet, between the Orange County bankruptcy and 2021, interest rates (across the yield curve) were fairly low and stable in the US. Why pay an asset/liability management team with expensive quants for 27 years to hedge an almost nonexistent risk that reappeared only in 2021? Even if quant departments do their job, will upper management listen to them?

On the other hand, history does not repeat itself. With the advent of financial derivatives and better regulation, markets have become more efficient. "Permanent regime changes sometimes occur, but they are rare" (Ang [6]). Pre-1933, the US yield curve was downward-sloping, whereas until 2021 it was upward-sloping. The implied volatility smile disappeared in US equities after the 1987 crash, to be replaced by a volatility smirk. A hefty profit could be gained in the 2010s from high-frequency trading. Since most large firms went into it, this profit has now eroded. An interesting trend today is the democratization of trading: "in January 2021, in the GameStop episode, retail investors coordinated using social media, resulting in the hedge fund Melvin Capital losing 53%" (Goldstein, Spatt, and Ye [55, section 4.3]), so even the conventional wisdom that sophisticated investors always win against small investors is being challenged.

Because of regime shifts, a sensible recommendation is to use only recent data to calibrate models. We thus tend to agree with Chen and Zhou [52, Chapter 1] that "compared with other domains where the amount of data is in the billions and trillions, the quantity of financial data is minuscule."

The popular literature focuses largely on the importance of big data but fails to acknowledge the difference between **big data** and **fast computers**. High-frequency traders do not process a lot of data but need fast computers. Firms that invest based on market sentiment analysis need large computers, to process a large quantity of poor-quality data, but not necessarily very fast ones. What about the others?

For investors who do not trade too frequently and whose trades do not move the market, we argue that they need neither big data nor large computers. A good knowledge of machine learning and asset pricing is sufficient. The algorithms we present in this book can be fairly computationally intensive but are small in terms of the quantity of code. See Chapter 3 in [52].

To do what? There is often money to be made or scientific knowledge to be developed when a domain is compartmentalized. At this point in time, many US asset managers do not use derivatives to fit their pricing models. Part of the difficulty comes from the fact that derivatives, as opposed to equities, are nonlinear instruments (observe, however, that this is an area where modern machine learning excels: nonlinearities). A better model of the SDF can be obtained if it is calibrated at the same time to a large cross-section of stocks and a large quantity of derivatives.[44] A caveat again for small investors: you are probably not going to win against Wall Street firms at their own game (their experience in pricing financial options is unmatched). However, not all options are traded. The large firms do not dominate real options. Pricing small- to medium-sized businesses or business opportunities (project finance) is a good area for smaller investors to excel. This is where, in our opinion, the democratization of finance may succeed. It is hard to see an episode like GameStop repeating itself: professional investors will rise or have already risen to the challenge of the democratization of trading. Small investors may want, as in the past, to think globally but act locally.

[44]There will need to be a harmonization in the methodologies. Financial engineers use parametric models that describe precisely the whole probability distributions, while asset managers look at sample moments (the first and the second one).

Of course, investors must be aware of current economic and political events. At the time of writing this book what struck us was

- *ESG investing* (environmental, social, and corporate governance): global warming and El Niño are modifying the earth climate
- high interest rates
- US government spending
- China real estate bubble
- the Ukraine–Russia conflict.

Chapter 16

Mean-Variance Portfolio Analysis

Mean-variance portfolio analysis is one of the triumphs of optimization applied to portfolio analysis. It was developed by Harry **Markowitz** in the 1950s [58], for which he obtained the Nobel Prize in economics. It started a whole new discipline, often called *modern portfolio theory*. In this chapter, we will see it applied to static portfolio optimization. The dynamic equivalent will be analyzed in Chapter 19 on *strategic asset allocation*.

As we shall see in Chapter 27 on empirical analysis, portfolios constructed using mean-variance analysis generally have poor performance and/or change considerably over time. Indeed, portfolio analysis is plagued by estimation problems on the mean and variance of the returns. The mean return of a security $E[r_i]$ is, in particular, difficult to estimate, as the assumption of ergodicity of returns[45] is a strong one. This book will cover two methodologies that attempt to alleviate this estimation problem. In the *Black–Litterman* approach, a Bayesian technique, one combines prior views with subjective views on the returns in the hope that returns will be more robustly estimated. In the *stochastic discount factor (SDF) methodology*, one approaches jointly the estimation (pricing) problem and the optimization (portfolio) problem.

The main contribution of this chapter resides thus more in its qualitative insights than in its results. Mean-variance analysis postulates that every optimal portfolio should be a linear combination of a risk-free investment and the *tangent portfolio,* which is composed of risky investments only. What distinguishes investors is the amount of money they put into the riskless asset versus what they put in the tangent portfolio. This leads to the *capital asset pricing model (CAPM)*, which shows that, in equilibrium, the tangent portfolio should be equal to the market portfolio.

The SDF methodology is more general than the CAPM in the sense that it is substantiated by weaker assumptions. It also delivers a portfolio construction methodology: it identifies the portfolios with the highest Sharpe ratio. An important result of this chapter (Theorem 16.4) shows indeed that the tangent portfolio (or, more generally, each portfolio on the capital market line) has optimal Sharpe ratio, thus establishing the link between mean-variance analysis and the SDF methodology. Again, this fact is important, as the latter, associated with machine learning, seems to deliver results superior to those of mean-variance analysis.

Mean-variance analysis studies the behavior of a single individual. We will see in Chapter 20 implications for the behavior of the market.

[45] Again, under that assumption, the average return is an unbiased estimator of the expected return. This fact is true under specific statistical assumptions of the return process. Random walks would qualify, or stationary processes, under the condition that they reached steady-state.

16.1 ▪ Definitions

We will consider only two types of assets: risky and riskless. Figure 16.1 shows returns of a risky series and returns of an (almost) riskless series.

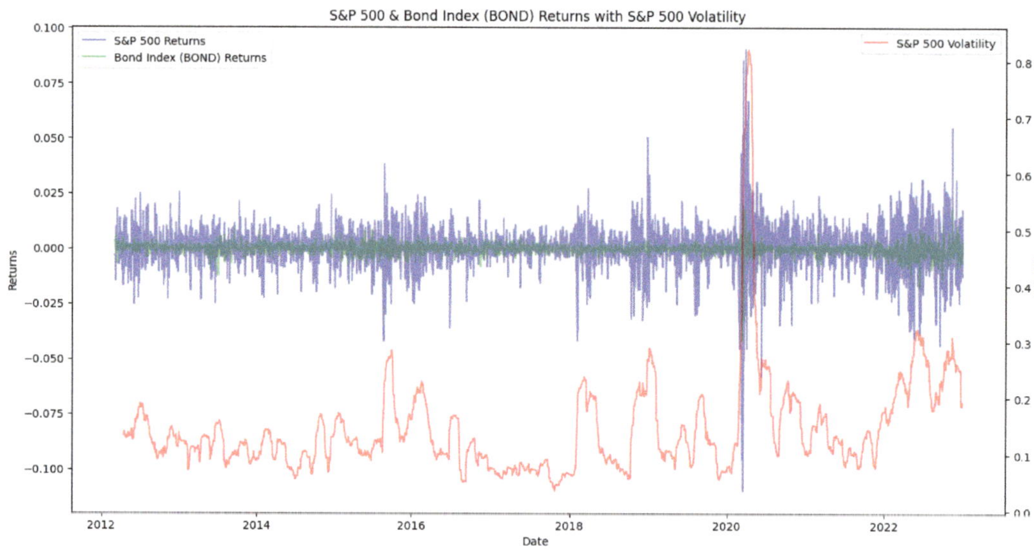

Figure 16.1: Time series of the S&P 500 return and volatility, and the bond index return time series.

An example of a risky asset is a **stock**. The owner of a stock of company i ($i = 1, \ldots, I$) receives at each time t a random *dividend* $d_{i,t}(\omega)$. Thus the payoff $x_{i,t+1}(\omega)$ of stock i at time $t+1$ (or price-cum-dividend) is equal to its *price ex-dividend* $p_{i,t+1}(\omega)$ plus the dividend:

$$x_{i,t+1}(\omega) = p_{i,t+1}(\omega) + d_{i,t+1}(\omega) \quad i \in \{1, 2, \ldots, I\}$$

The **riskless asset** is the one-period Treasury bond (or T-bill). It is numbered security $i = 0$. For a one-period Treasury bond issued at time t, the payoff $x_{0,t+1}$ will be nonrandom and, by convention in this book, equal to the principal ($1):

$$x_{0,t+1} = p_{0,t+1} = 1$$

Economists and portfolio managers find it more intuitive to talk about returns than about prices. Also, net returns are more likely to be normally distributed than prices. Gross (arithmetic) returns are simply future payoff divided by current (ex-dividend) prices. Today's (time t) prices are known, but not tomorrow's (time $t+1$) prices, and to emphasize this fact, we drop the ω symbol from $p_{i,t}(\omega)$:

$$R_{i,t+1}(\omega) = \frac{x_{i,t+1}(\omega)}{p_{i,t}}$$

Net (arithmetic) returns are gross returns minus one. The notation is summarized in Table 16.1.

We will concatenate the net returns for risky securities into a net return vector:

$$\mathbf{r}_{t+1}(\omega) = \begin{bmatrix} r_{1,t+1}(\omega) \\ \vdots \\ r_{I,t+1}(\omega) \end{bmatrix}$$

Notation	Name	Definition
$R_{i,t+1}$	Gross return of stock i	$(d_{i,t+1} + p_{i,t+1})/p_{i,t}$
$r_{i,t+1}$	Net return of stock i	$R_{i,t+1} - 1$
$R_{f,t}$	Gross return of risk-free bond	$\frac{1}{p_{0,t}}$
$r_{f,t}$	Net return of a risk-free bond	$R_{f,t} - 1$
$r^e_{i,t}$	Excess return of stock i	$r^e_{i,t} - r_{f,t}$

Table 16.1: Notation for returns.

A portfolio at time t consists of a number of shares $(n_{0,t}, n_{1,t}, \ldots, n_{I,t})$ of each security. The value of the portfolio ex-dividends, or *wealth*, at time t is denoted by $W(t)$ and is equal to

$$W(t) = \sum_{i=0}^{I} n_{i,t} p_{i,t}$$

One could also define (but rarely use) the value of the portfolio cum-dividends:

$$W^c(t+1,\omega) = \sum_{i=0}^{I} n_{i,t} x_{i,t+1}(\omega)$$

The rationale for these two definitions is that we make our stock allocation (at time t) after dividends are paid.[46] For reasons that will become clear shortly, it is often more convenient to state the portfolio problem in terms of return than in terms of dollar values. To that effect, we define the *net portfolio return* as

$$r_{p,t+1}(\omega) = \frac{W^c(t+1,\omega)}{W(t)} - 1$$

We now define the *portfolio weight* $w_{i,t}$ as the proportion of the portfolio wealth invested in each security, i.e.,

$$w_{i,t} = \frac{n_{i,t} p_{i,t}}{W_t}$$

With simple arithmetic, we obtain

$$\begin{aligned} r_{p,t+1}(\omega) &= \frac{\sum_{i=0}^{I} n_{i,t} x_{i,t+1}(\omega)}{W_t} - 1 \\ &= \frac{\sum_{i=0}^{I} n_{i,t} p_{i,t} \frac{x_{i,t+1}(\omega)}{p_{i,t}}}{W_t} - 1 \\ &= \sum_{i=0}^{I} w_{i,t} R_{i,t+1}(\omega) - 1 \\ &= \sum_{i=0}^{I} w_{i,t} r_{i,t+1}(\omega) \end{aligned}$$

The last equation is easily obtained since the portfolio weights add up to one. Defining the following two vectors:

$$\mathbf{w}_t = \begin{bmatrix} w_{1,t} \\ \vdots \\ w_{I,t} \end{bmatrix} \quad \mathbf{1} = \begin{bmatrix} 1 \\ \vdots \\ 1 \end{bmatrix}$$

[46] The reader may wonder where the dividends in period t went in this definition. This is a concern only in multiperiod problems and will be seen in strategic allocation in Chapter 19.

We thus write the *budget constraint*:

$$1 = w_{0,t} + \mathbf{1}'\mathbf{w}_t \tag{16.1}$$

Thus, one obtains the following (trivial yet important) fact.

Fact 16.1. *It is equivalent to maximize at time t (a measure of) future wealth $W^c(t+1)$ over the number of shares, under a constraint that the portfolio can be financed, and to maximize $r_{p,t+1}$ under the self-financing constraint (16.1).*

The investor cares about two measures: **expected return** and **risk**. We denote the conditional mean of the net stock return (in short, *expected stock return*) by

$$\bar{\mathbf{r}}_t = E_t[\mathbf{r}_{t+1}]$$

The portfolio expected return $\bar{r}_{p,t}$ is then simply equal to $w_{0,t}r_{f,t} + \bar{\mathbf{r}}_t'\mathbf{w}_t$.

While there are many different measures of risk, the simplest one is the variance of return. We denote by $\mathbf{\Sigma}_t(i,j)$ the conditional covariance at time t between $r_{i,t+1}$ and $r_{j,t+1}$.

$$\mathbf{\Sigma}_t(i,j) = Cov_t[r_{i,t+1}, r_{j,t+1}]$$

Suppose we have two assets ($I = 2$). Let us drop for a moment the time subscript. The variance of the return of the portfolio $r_p(\omega) = w_1 r_1(\omega) + w_2 r_2(\omega)$ (denoted by σ_p^2) is given by

$$\begin{aligned}\sigma_p^2 &\equiv Var[w_1 r_1 + w_2 r_2] = Var[w_1 r_1] + Var[w_2 r_2] \\ &\quad + 2Cov[w_1 r_1, w_2 r_2] \\ &= w_1^2 \Sigma_{1,1} + 2 w_1 w_2 \Sigma_{1,2} + w_2^2 \Sigma_{2,2}\end{aligned}$$

For more than two assets, we have the same rule. The variance of the return on the portfolio is

$$\sigma_{p,t}^2 = \mathbf{w}'\mathbf{\Sigma}_t\mathbf{w}. \tag{16.2}$$

16.2 ▪ Conditional versus Unconditional Mean-Variance

While the optimality conditions we develop obviously hold regardless of the definition of the horizon (here we chose a horizon of one time step), it is critical to remember that the numerical solutions to the portfolio problem can be potentially very different if we optimize the portfolio over one period (t to $t+1$) or over the full horizon (0 to T). In the first case, we work with **conditional** expected/variance of the return, while in the second case, we work with **unconditional** expected/variance of the return, i.e., $\bar{r}_i = E[\frac{p_{i,t+1}}{p_{i,0}}] - 1$, and $\Sigma(i,j) = Cov[r_{i,t+1}, r_{j,t+1}]$. The results will thus vary considerably depending on the horizon. This difference between these two types of moments is so important that we decided to catch the attention of the reader by putting it in a figure, namely, Figure 16.2.

We consider in this chapter the static problem, where the investor chooses at time zero an optimal portfolio vector at time zero, namely, $\mathbf{w}_0^*$, which will be held constant over the horizon (0 to T).[47] *Strategic asset allocation*, also called *dynamic portfolio choice*, is the methodology where, while optimizing over the same horizon (0 to T), one considers the possibility of rebalancing the portfolio at every time step $\{0, 1, 2, \ldots, T-1\}$, that is, modifying the portfolios over the horizon. One will then construct at time zero an optimal portfolio:

$$\{\mathbf{w}_0^*, \mathbf{w}_1^*(\omega), \ldots, \mathbf{w}_{T-1}^*(\omega)\}$$

[47] Of course the investor is still free to change his plan at any time, say time 1, based on new information. If there is no new information, however, the optimal plan between time 0 and T is the same as the optimal plan between time 1 and $T+1$.

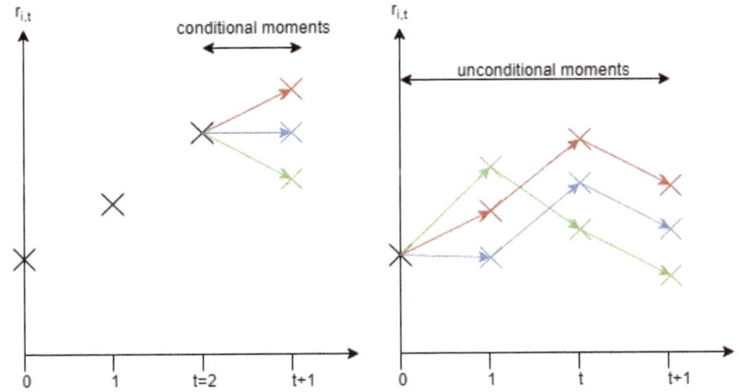

Figure 16.2: Three scenarios (red, blue, and green) of returns. To the left, the "conditional" mean-variance problem: investors optimize their portfolio at time t for horizon $t+1$. To the right, the "unconditional" mean-variance problem: investors optimize their portfolio at time 0 for horizon $t+1$.

Observe that future optimal portfolios will depend on the state of the world ω. In theory, dynamic portfolio selection will be superior to static portfolio allocation since it has more decision variables. In practice, it is made more difficult by estimation problems, as it will need to estimate not only the unconditional moments but also the conditional moments, a harder task. Dynamic portfolio selection will be covered in Chapter 19. We now use the following **convention** in this chapter: we drop the time subscript $_t$ on all our variables.

16.3 ▪ The Mean-Variance Problem without a Riskless Asset

While asset prices are obviously not normally distributed, net returns tend to be, especially when the horizon is very small.[48] When the horizon is large, the normal assumption is frequently rejected, and investors must take into account higher moments such as **skewness** (third standardized moment) and **kurtosis** (fourth standardized moment). We will not consider this case here. Since the Gaussian distribution is characterized by only two parameters, the mean and the variance, it is natural to formulate the portfolio problem in terms of these parameters and in terms of net returns (as opposed to wealth).

A risk-averse investor would like to maximize the expected portfolio return $\bar{r}_p$ while minimizing the portfolio variance σ_p^2. This is an example of *goal programming*, where a decision-maker has more than one objective.

Markowitz's approach is to fix one of the parameters and optimize the other parameter. Thus the solution is a family of solutions indexed by the fixed value of one of the two parameters. Markowitz fixed the required expected return, i.e., $\bar{r}_p = \mu$, and minimized the variance. It should be clear (see Exercise 4.8) that one obtains the same family of solutions if one fixes the variance and maximizes the expected return. One obtains deep insight into the portfolio problem by looking at the geometric properties of the minimum variance set, i.e., the set of optimal variances for a given μ; see Figure 16.3.

Without a riskless asset , we have the constraint

$$w_0 = 0$$

[48]When the horizon tends to zero, every net return becomes Gaussian, with the conditional variance potentially stochastic. Thus any return distribution can be reproduced with stochastic volatility. This is a consequence of the martingale representation theorem in stochastic calculus.

Thus the expected portfolio net return simplifies into

$$\bar{r}_p = \bar{\mathbf{r}}'\mathbf{w}$$

The *mean-variance problem* is to minimize the portfolio variance, given by (16.2):

$$\min_{\mathbf{w}} \frac{1}{2}\mathbf{w}'\boldsymbol{\Sigma}\mathbf{w}$$

$$\mathbf{1}'\mathbf{w} = 1 \tag{16.3}$$

$$\bar{\mathbf{r}}'\mathbf{w} = \mu \tag{16.4}$$

We have not imposed positivity constraints of the form $w_i \geq 0$ so that unrestricted short sales are permitted. We define the Lagrangian:

$$L(\mathbf{w},\lambda,\gamma) = \frac{1}{2}\mathbf{w}'\boldsymbol{\Sigma}\mathbf{w}+\lambda(1-\mathbf{1}'\mathbf{w}) + \gamma(\mu - \bar{\mathbf{r}}'\mathbf{w})$$

The first-order conditions for the optimal portfolio $\mathbf{w}^*$ are

$$\boldsymbol{\Sigma}\mathbf{w}^* - \lambda\mathbf{1} - \gamma\bar{\mathbf{r}} = \mathbf{0} \tag{16.5}$$

An important operation in mean-variance analysis is inverting the return variance-covariance matrix $\boldsymbol{\Sigma}$.[49] Thus, premultiplying (16.5) by $\boldsymbol{\Sigma}^{-1}$, the optimal portfolio is

$$\mathbf{w}^* = \lambda\boldsymbol{\Sigma}^{-1}\mathbf{1} + \gamma\boldsymbol{\Sigma}^{-1}\bar{\mathbf{r}} \tag{16.6}$$

Now the equality constraints (16.3) and (16.4) must be satisfied. Introducing (16.6) into (16.3), (16.4) gives

$$\begin{bmatrix} A & B \\ B & C \end{bmatrix} \begin{bmatrix} \lambda \\ \gamma \end{bmatrix} = \begin{bmatrix} 1 \\ \mu \end{bmatrix} \tag{16.7}$$

where we defined the following constants:

$$A = \mathbf{1}'\boldsymbol{\Sigma}^{-1}\mathbf{1}$$

$$B = \mathbf{1}'\boldsymbol{\Sigma}^{-1}\bar{\mathbf{r}}$$

$$C = \bar{\mathbf{r}}'\boldsymbol{\Sigma}^{-1}\bar{\mathbf{r}}$$

Since $\boldsymbol{\Sigma}$ is positive definite, so is $\boldsymbol{\Sigma}^{-1}$. We can rapidly infer that $A \geq 0, C \geq 0$. Define the determinant $\Delta = AC - B^2$. By the Cauchy–Schwarz inequality, we can see that $\Delta \geq 0$. When $\Delta > 0$,

$$\begin{bmatrix} \lambda \\ \gamma \end{bmatrix} = \frac{1}{\Delta} \begin{bmatrix} C & -B \\ -B & A \end{bmatrix} \begin{bmatrix} 1 \\ \mu \end{bmatrix} = \begin{bmatrix} \frac{C-\mu B}{\Delta} \\ \frac{\mu A - B}{\Delta} \end{bmatrix} \tag{16.8}$$

The Lagrange multipliers are a function of the required return μ. For any value of μ there will be an optimal portfolio $\mathbf{w}^*(\mu)$. We turn now to characterizing this set, called the *minimum variance frontier*.

[49]We will see that one of the advantages of using machine learning (in particular, the stochastic gradient descent method) is that we bypass this operation, which is often numerically delicate.

16.3.1 ▪ Minimum Variance Frontier

We now express[50] the optimal portfolio return variance $\sigma^2 = \mathbf{w}^{*\prime}\mathbf{\Sigma}\mathbf{w}^*$ as a function of its expected return μ. Using (16.6) and (16.8), we obtain

$$\begin{aligned}\sigma^2 &= \mathbf{w}^{*\prime}\mathbf{\Sigma}\mathbf{w}^* = \mathbf{w}^{*\prime}\mathbf{\Sigma}(\lambda\mathbf{\Sigma}^{-1}\mathbf{1} + \gamma\mathbf{\Sigma}^{-1}\bar{\mathbf{r}})\\ &= \lambda\mathbf{w}^{*\prime}\mathbf{1}+\gamma\mathbf{w}^{*\prime}\bar{\mathbf{r}}\\ &= \lambda + \gamma\mu\\ &= \frac{C-\mu B}{\Delta} + \frac{\mu A - B}{\Delta}\mu\\ &= \frac{A\mu^2 - 2B\mu + C}{\Delta}\end{aligned}$$

This is the equation of a parabola. The minimizer μ^* of this parabola is attained at

$$\frac{d\sigma^2}{d\mu}|\mu^* = 0$$

We obtain

$$\mu^* = \frac{B}{A}$$

Usually, investors obtain positive excess returns when investing in stocks, which enables us to confirm that, since $A \geq 0$, we should have $B \geq 0$.

Despite its name, the minimum variance frontier is usually represented in mean return/standard deviation space and not in mean return/variance space; thus we obtain

$$\sigma = \sqrt{\frac{A\mu^2 - 2B\mu + C}{\Delta}}$$

This is the equation of a hyperbola, and, in the nondegenerate case, we obtain Figure 16.3. The slope of the asymptotes to the minimum variance frontier can be obtained by finding $\frac{d\mu}{d\sigma}|_{\sigma=\infty}$.

$$\begin{aligned}\frac{d\mu}{d\sigma}|_{\sigma=\infty} &= \frac{d\mu}{d\sigma^2}\frac{d\sigma^2}{d\sigma}|_{\sigma=\infty}\\ &= \frac{2\sigma}{d\sigma^2/d\mu}|_{\mu=\infty}\\ &= \frac{\sigma\Delta}{A\mu - B}|_{\mu=\infty}\\ &= \frac{\sqrt{\frac{A\mu^2-2B\mu+C}{\Delta}}\Delta}{A\mu - B}|_{\mu=\infty}\\ &= \sqrt{\frac{\Delta}{A}}\end{aligned}$$

The upper part of the minimum variance frontier is called the **mean-variance efficient frontier** (see Figure 16.3). It is clear that a risk-averse investor should only choose portfolios on the mean-variance efficient frontier.

[50]We drop the subscript p on σ_p^2 as is the convention.

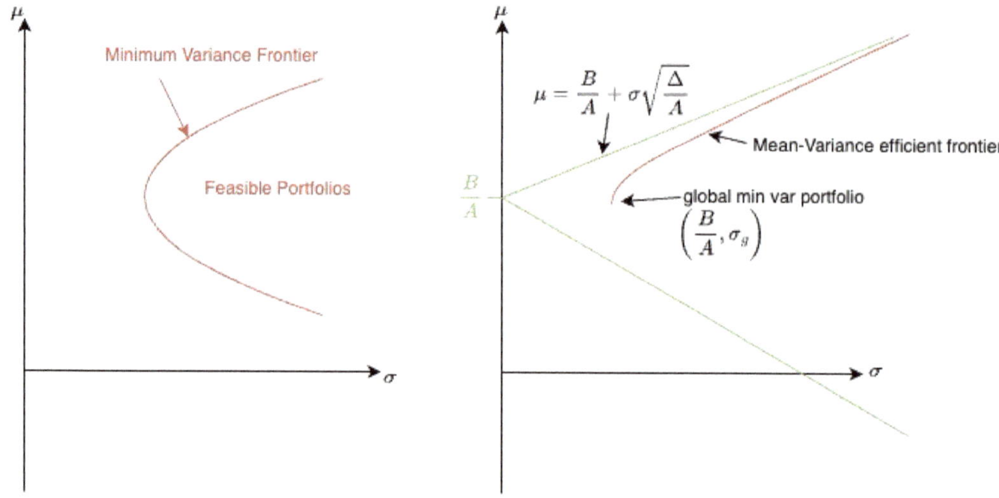

Figure 16.3: Mean-variance frontier without a riskless asset.

16.4 ▪ Two-Fund Separation

Two-fund separation is barely more than equation (16.6) and recalling that, by definition, a **fund** or **stock portfolio** is a weight vector $\mathbf{w}$ whose coefficients add up to one. From equation (16.8), we have

$$\lambda = \frac{C - \mu B}{\Delta}$$
$$\gamma = \frac{\mu A - B}{\Delta}$$

Thus,

$$\lambda A + \gamma B (= \frac{C - \mu B}{\Delta} A + \frac{\mu A - B}{\Delta} B) = 1 \tag{16.9}$$

The following lines show that any mean-variance efficient portfolio is the sum of two portfolios ("funds"), i.e., linear combinations of securities that add up to one. From (16.6),

$$\mathbf{w}^* = \lambda \boldsymbol{\Sigma}^{-1}\mathbf{1} + \gamma \boldsymbol{\Sigma}^{-1}\bar{\mathbf{r}} \tag{16.10}$$
$$= \lambda A \frac{\boldsymbol{\Sigma}^{-1}\mathbf{1}}{A} + \gamma B \frac{\boldsymbol{\Sigma}^{-1}\bar{\mathbf{r}}}{B} \tag{16.11}$$

By (16.9), λA and γB are coefficients of a convex combination (i.e., they sum up to one). We define

$$\mathbf{w}_g = \frac{\boldsymbol{\Sigma}^{-1}\mathbf{1}}{A}$$
$$\mathbf{w}_d = \frac{\boldsymbol{\Sigma}^{-1}\bar{\mathbf{r}}}{B}$$

By construction, both $\mathbf{w}_g$ and $\mathbf{w}_d$ are *portfolios*, i.e., $\mathbf{w}_g'\mathbf{1} = \mathbf{w}_d'\mathbf{1} = 1$, so that $\mathbf{w}^*$ is made of investing a proportion λA of one's wealth in portfolio g and a proportion γB in portfolio d. As you will see in the exercises, $\mathbf{w}_g$ is the minimum variance portfolio.

16.5 ▪ The Portfolio Problem with a Riskless Asset

Recall that asset zero is the risk-free bond. The investor now invests a fraction $w_0 \neq 0$ of his/her wealth in the riskless asset, so we use our original budget constraint (16.1). In what follows, short-selling and borrowing are allowed so that the decision variables will only be $\mathbf{w}$, and $\mathbf{w}$ will be unrestricted. The budget constraint (16.1) will be automatically satisfied once we calculate $\mathbf{w}^*$ by setting the optimal portfolio wealth to be

$$w_0^* = 1 - \mathbf{1}'\mathbf{w}^*$$

We now extend the expected return constraint (16.4) to include the risk-free asset:

$$\begin{aligned}\mu &= r_f w_0 + \bar{\mathbf{r}}'\mathbf{w} \\ &= r_f(1 - \mathbf{1}'\mathbf{w}) + \bar{\mathbf{r}}'\mathbf{w} \\ &= r_f + (\bar{\mathbf{r}} - r_f\mathbf{1})'\mathbf{w}\end{aligned}$$

The quantity $\bar{\mathbf{r}} - r_f\mathbf{1}$ is called the **excess** (compared to the risk-free rate) **expected return** and plays a fundamental role in asset pricing. We can now write our optimization problem as

$$\min_{\mathbf{w}} \frac{1}{2}\mathbf{w}'\mathbf{\Sigma}\mathbf{w} \tag{16.12}$$

$$(\bar{\mathbf{r}} - r_f\mathbf{1})'\mathbf{w} = \mu - r_f \tag{16.13}$$

Associating the Lagrange multiplier γ to the constraint, we have

$$\mathbf{\Sigma}\mathbf{w}^* - \gamma(\bar{\mathbf{r}} - r_f\mathbf{1}) = \mathbf{0}$$

We obtain thus one of the most important results of this book. To highlight it, we make it a theorem.

Theorem 16.2. *The **optimal portfolio** is*

$$\mathbf{w}^* = \gamma\mathbf{\Sigma}^{-1}(\bar{\mathbf{r}} - r_f\mathbf{1}) \tag{16.14}$$

To calculate γ, we reintroduce it in the constraint (16.13) and obtain

$$\gamma = \frac{\mu - r_f}{(\bar{\mathbf{r}} - r_f\mathbf{1})'\mathbf{\Sigma}^{-1}(\bar{\mathbf{r}} - r_f\mathbf{1})} \tag{16.15}$$

In other words, the magnitude of the stock portfolio (16.14) is proportional[51] to the excess return $\mu - r_f$. In general, $\mu > r_f$, so the higher the required expected return of the investor, the larger the stock portfolio should be. We can also reintroduce (16.14) into the optimal portfolio return variance $\sigma^2 = \mathbf{w}^{*\prime}\mathbf{\Sigma}\mathbf{w}^*$ and obtain

$$\mu - r_f = \pm\sigma\sqrt{C - 2r_fB + r_f^2A}$$

This is the equation of two straight lines in mean-variance space. The top one is called either the **capital (market) line** or the **global efficient frontier** (see Figure 16.4). The capital market line is tangent to the mean-variance efficient frontier. The point where they intersect is (the mean and standard deviation of) the **tangency portfolio** (return). You will prove in the exercises that

[51]The denominator depends on the market, but not on the preference (μ) of the investor.

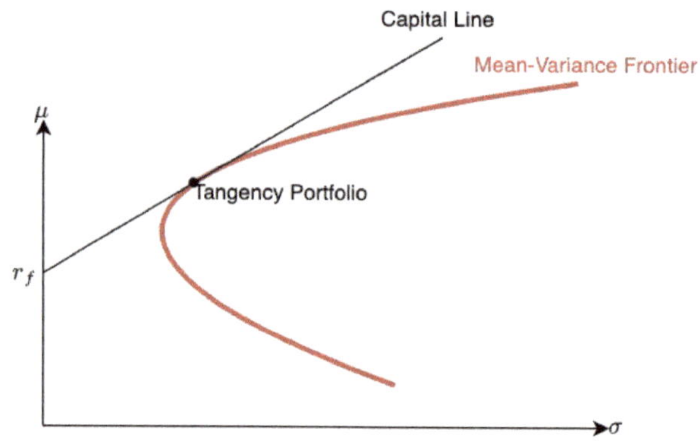

Figure 16.4: Capital market line (a.k.a. global efficient frontier), mean-variance frontier, and tangency portfolio.

the tangency portfolio, defined by

$$\mathbf{w}^T = \frac{\boldsymbol{\Sigma}^{-1}(\bar{\mathbf{r}}-r_f\mathbf{1})}{B - Ar_f} \tag{16.16}$$

is indeed both on the minimum variance frontier and on the capital market line.

Definition 16.3. *The Sharpe ratio of security i is its expected excess return divided by its volatility:*

$$SR_i = \frac{E[r_i] - r_f}{\sqrt{Var[r_i]}}$$

Take a point on the capital market line. Any asset is "to the right," that is, for the same mean, has higher volatility. Thus any feasible portfolio has a Sharpe ratio that is smaller than or equal to a portfolio on the capital market line.

Theorem 16.4. *Any portfolio on the global efficient frontier has maximum Sharpe ratio.*

Exercises

16.1. (Easy) Use the following data: $r_f = 0$,

$$\bar{\mathbf{r}} = \begin{bmatrix} 0.2 \\ 0.1 \\ -0.1 \end{bmatrix} \quad \boldsymbol{\Sigma} = \begin{bmatrix} 0.2 & 0.1 & 0.05 \\ 0.1 & 0.4 & 0 \\ 0.05 & 0 & 0.6 \end{bmatrix}$$

(a) Calculate $\mathbf{w}_g$ and $\mathbf{w}_d$, as well as the mean return and variances of these portfolios.

(b) Calculate the tangent portfolio $\mathbf{w}^T$ as well as the corresponding mean and variance.

(c) Now add the no-short sales constraint:

$$\mathbf{w} \geq 0.$$

Solve numerically for the optimal portfolio corresponding to the mean return $\mu = 1$. Feel free to use Excel or MATLAB.

16.2. (Easy) Show that $\mathbf{w}_g$ is the global minimum variance portfolio.

16.3. (Easy) Show that $\boldsymbol{\Sigma}_t$ is also the variance-covariance matrix of the gross return and of the excess return.

16.4. (Medium) Use well-known probability facts to show that any variance-covariance matrix is positive definite. Use spectral decomposition to show that its inverse is also positive definite.

16.5. (Medium) Show that A, C, and Δ are positive.

16.6. (Medium) Consider I firms. The returns on their stocks are IID with mean m and variance v. What is Δ? Describe the minimum variance set. What happens when $I \to \infty$?

16.7. (Medium) Prove Theorem 16.4 algebraically (if needed, by specifying further assumptions) by

(a) showing that the tangency portfolio is a member of the mean-variance set;

(b) calculating the slope of the capital line;

(c) showing that the capital market line is tangent to the mean-variance set at the tangency portfolio, and then conclude. (We use the ubiquitous term "tangency portfolio" to refer to the mean and standard deviation of the tangency portfolio return.)

16.8. (Hard) Consider the Markowitz problem with a risk-free asset. Let us parameterize the minimal portfolio variance σ_p^2 as well as the optimal portfolio $\mathbf{w}^*$ by excess return $\mu - r_f$; i.e., σ_p^2 and $\mathbf{w}^*$ are functions of $\mu - R$. Suppose that

$$\mathbf{w}^*(0.02) = 2\boldsymbol{\Sigma}^{-1}(\bar{\mathbf{r}} - r_f\mathbf{1})$$

Suppose that we increase the required excess return by 0.01. How much do we increase σ_p^2?

Hint: Remember the interpretation of the Lagrange multipliers from Chapter 3.

16.9. (Hard) In this problem you will add a Value-at-Risk (VaR) constraint. Consider the following mean-variance problem, to which we add a VaR constraint.
Define $Var_c[r_p]$ as the minimum (in general) positive numerical value such that

$$P(r_p \geq -Var_c[r_p]) = c$$

In other terms, it is the quantile of the distribution of (minus) the portfolio return $r_p(\omega)$. The problem is

$$\min \mathbf{w}^t \Sigma \mathbf{w}$$
$$\bar{\mathbf{r}}^t \mathbf{w} = \mu - r_f$$
$$Var_c[r_p] \leq \gamma$$

where γ and c are percentages with the following interpretation. Suppose $\gamma = 10\%$ and $c = 95\%$. The constraint $Var_c[r_p] \leq \gamma$ means that with 95% confidence the loss on the portfolio is less than 10% of the value of the portfolio. Assume that returns are normally distributed.

(a) Give an expression for $Var_c[r_p]$ as a function of $\mathbf{w}^t\Sigma\mathbf{w}$,$\mu - r_f$, c, and the inverse of the normal distribution N^{-1}.

(b) Represent graphically the VaR constraint in mean /standard deviation space.

(c) Determine graphically the maximal value of μ.

(d) Suppose now that the portfolio is not necessarily normally distributed. How should you modify the VaR constraint? (Hint: Use Chebyshev's inequality.)

Chapter 17

Expected Utility Theory

One of the main ingredients of financial economics is modeling agent preferences. Traditionally, one models agents as risk-averse. In behavioral economics, this constraint is relaxed. We start with risk-averse agents.

17.1 ▪ Risk-Averse Agents

We model risk-averse investors' preferences with concave and increasing *utility functions*. An agent that maximizes the expected utility of (future) wealth, with such a utility function, prefers[52]

(1) more wealth to less wealth, because the utility function is increasing

(2) certain wealth to uncertain wealth because the utility function is concave.

These utility functions are called *von Neumann–Morgenstern* utility functions.

To prove the second point, suppose that an investor has the choice between two investments with the same expected return. Investment 1 is the *riskless* one, and investment 2 is the *risky* one. They both pay one period from now. Denote by $x_1 \in \mathbb{R}$ the constant payoff of investment 1 in the future, and by $x_2(\omega) \in \mathbb{R}$ the random payoff of investment 2 in the future. They will have the same expected value if $E[x_1] = E[x_2]$. The first investment is the safe one, meaning that it is not random:

$$x_1 = E[x_1]$$

while the second one is random, i.e., $x_2(\omega) \neq E[x_2]$.

Theorem 17.1. *If $u : \mathbb{R} \to \mathbb{R}$ is concave increasing, and if x_1 is riskless and x_2 risky, with $x_1 = E[x_2]$, then*

$$E[u(x_1)] \geq E[u(x_2)]$$

Thus an agent who maximizes expected utility will prefer investment 1 to investment 2.

An illustration of the proof is provided below in Figure 17.1.

Proof. Since x_1 is not random,

$$E[u(x_1)] = u(x_1) = u(E[x_1]) \tag{17.1}$$

[52]The third criterion is that investors prefer to receive cash now than cash tomorrow. We will use that fact in a moment.

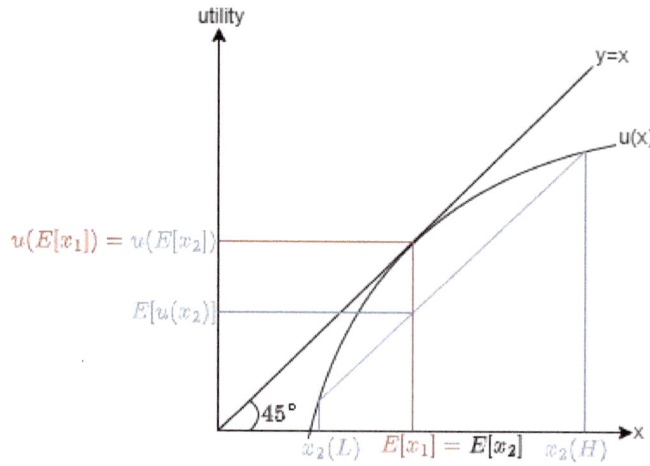

Figure 17.1: Illustration of Jensen's inequality. Investment 1 is nonrandom, with payoff x_1. Investment 2 has two outcomes: $x_2(L)$ and $x_2(H)$ such that $E[x_2] = E[x_1] = x_1$.

By concavity of u and Jensen's inequality,

$$E[u(x_2)] \leq u(E[x_2]) \tag{17.2}$$

Since the two investments have the same mean, i.e., $E[x_1] = E[x_2]$, then

$$u(E[x_2]) = u(E[x_1]) \tag{17.3}$$

Assembling (17.1), (17.2), and (17.3), we have

$$\begin{aligned} E[u(x_1)] &= u(E[x_1]) \\ &= u(E[x_2]) \\ &\geq E[u(x_2)] \end{aligned}$$

■

Typical examples of utility functions are

- the exponential $u(x) = -\exp(-ax)$
- the logarithm $u(x) = \ln(x)$
- the power function $u(x) = x^\alpha$ for $0 < \alpha < 1$.

We also saw in Chapter 5 that concave utility functions can be approximated by piecewise linear functions. The usefulness of this approximation stems from the fact that the optimal investment under a budget constraint can be found by linear programming. This approximation does not hold for utility functions that are not concave, to which we now turn.

17.2 ▪ Behavioral Economics

Tourists who gamble in Las Vegas are clearly not risk-neutral but risk-seeking. Risk aversion can also depend on the level of wealth of an individual. Thus, not everybody models investors with von Neumann–Morgenstern [59] utility functions.

In *prospect theory*, developed by Kahneman and Tversky [60],

- investors find the pain of losses greater than the joy of gains; loss aversion utility allows the investor to have different marginal utilities for gains and losses
- probabilities in the expected utility calculation can be very subjective.

Or one could use a utility function that is convex (and thus risk-seeking) over losses but concave over gains.

In machine learning, one rarely uses these functions. The danger of the gradient algorithms not converging is magnified by the lack of concavity of these functions.

Exercises

17.1. (Medium) Suppose an investor can make at most $2000 and at least $0 over one day. He is indifferent to

- making $2000 with probability 1/2 or making $800 for sure
- making $800 with probability 1/2 or making $300 for sure
- making $2000 with probability 1/2 and $800 with probability 1/2 or making $1000 for sure.

The investor's utility function $u(x)$ is piecewise linear. Construct it recursively.
Hint: Set $u(0) = 0$, $u(2000) = 1$. Using the first condition, we find, for $x = 800$,

$$u(x) = \frac{1}{2}u(2000) + \frac{1}{2}u(0)$$

17.2. Suppose you have the choice between two stocks, with the following payoff:

	$\omega = \omega_1$	$\omega = \omega_2$	$\omega = \omega_3$
$x_1(\omega)$	100	120	0
$x_2(\omega)$	0	170	60

Each state $\omega = \omega_s$ occurs with probability 1/3. One invests n_1 in stock x_1 and n_2 in stock x_2. The price of stock 1 is $95, while the price of stock 2 is $100. The investor has a budget of $1500. The utility function is the one from Exercise 17.1.

(a) Formulate the problem as a linear program.

(b) Solve it, using either Excel or MATLAB.

Chapter 18
The Black–Litterman Approach

Notation: In this chapter, all random variables are denoted with a tilde to distinguish them from their realization; for instance, $\tilde{\mathbf{r}}$ is the return viewed as a random variable, and $\mathbf{r} \equiv \tilde{\mathbf{r}}(\omega)$ its realization.

Black and Litterman [61] pioneered the Bayesian approach to modern portfolio theory. As we saw in Chapter 16 on mean-variance, the parameters of the model are very uncertain. The Black–Litterman approach acknowledges this uncertainty by postulating that the expected return $\tilde{\boldsymbol{\mu}}(\omega)$ is a random variable.

The model for the return is:

$$\tilde{\mathbf{r}}(\omega) = \tilde{\boldsymbol{\mu}}(\omega) + \tilde{\mathbf{u}}(\omega)$$

where the mean-zero normal random variable $\tilde{\mathbf{u}}$ represents market risk. We write $Var[\tilde{\mathbf{u}}] = \boldsymbol{\Sigma}$. Thus, if $\tilde{\boldsymbol{\mu}}$ is not random, we are back to mean-variance. We use the notation $\bar{\mathbf{r}} = E[\tilde{\mathbf{r}}]$.

There are thus two sources of randomness in the model: traditional market risk and model risk (sometimes called ambiguity of the parameters), as can be seen from the formula of the variance:

$$Var[\tilde{\mathbf{r}}] = \underbrace{Var[E[\tilde{\mathbf{r}}|\tilde{\boldsymbol{\mu}}]]}_{\text{ambiguity}} + \underbrace{E[Var[\tilde{\mathbf{r}}|\tilde{\boldsymbol{\mu}}]]}_{\text{market risk}} \tag{18.1}$$

Usually $\tilde{\mathbf{u}}$ is assumed independent from $\tilde{\boldsymbol{\mu}}$ so that the market risk variance-covariance matrix is equal to $\boldsymbol{\Sigma}$. Let us now focus on the ambiguity term.

18.1 ▪ Motivation

The Black–Litterman (BL) approach solves several problems that practitioners face when applying mean-variance portfolio allocation (see Chapter 16):

1. the tangent portfolio turns out to be vastly different from the equilibrium (market) portfolio[53]
2. ambiguity adds to the risk of investing, and this ambiguity should be recognized
3. investors often have **views** on a limited number of assets; how to incorporate the views in a way that satisfies their preferences?
4. the portfolio allocations resulting from mean-variance analysis turn out not to be robust.

[53]That is, the CAPM does not hold.

Greater robustness of the BL approach can be seen in (18.1). An estimator of $E[Var[\tilde{\mathbf{r}}|\tilde{\mu}]]$ should be more precise than an estimator of $Var[\tilde{\mathbf{r}}]$. We illustrate the first point with an example.

Example 18.1. We define the *equilibrium portfolio* $\mathbf{w}^M$ as the ratio of the market capitalization of each asset class to the total value of the assets in the world.

We take three different (risky) asset classes: European assets, American assets, and Asian assets. The sum of them is (approximately) equal to the world portfolio, and each asset class is approximately equal in value. Thus we expect to observe the following proportion of each continent in the global portfolio:

$$\mathbf{w}^M = \begin{bmatrix} \text{Europe proportion} \\ \text{US proportion} \\ \text{Asia proportion} \end{bmatrix} \simeq \begin{bmatrix} 1/3 \\ 1/3 \\ 1/3 \end{bmatrix}$$

We calculated the tangency portfolio $\mathbf{w}^T$ in 2015 using simple estimation methods for the asset expected returns μ and variance $\mathbf{\Sigma}$ and obtained (with A and B defined as in Chapter 16)

$$\mathbf{w}^T = \frac{\mathbf{\Sigma}^{-1}(\bar{\mathbf{r}} - r_f\mathbf{1})}{B - Ar_f} \simeq \begin{bmatrix} -3.5 \\ -0.1 \\ 3.6 \end{bmatrix}$$

If the market were in a CAPM-like equilibrium and if the parameter estimation were perfect, we would have

$$\mathbf{w}^M = \mathbf{w}^T$$

as will be proved in Chapter 20. Obviously, this was not the case in 2015.

18.1.1 ▪ Views on Asset Returns

In practice, many investors have simple beliefs, such as

- I believe in the market
- I want to invest in this sector, which I know best
- I want to invest only in my country
- I want to do pair trading.

How do we incorporate these beliefs? As an example, let

$$\mathbf{r} = \begin{bmatrix} r_1 \\ r_2 \end{bmatrix} = \begin{bmatrix} \text{Germany's return} \\ \text{France's return} \end{bmatrix}$$

Suppose that we believe that Germany will do 5% better than France. We could try to put in the model

$$r_1 - r_2 = 5\%$$

This is **not** how we will incorporate views. In the BL approach, views will be taken on *expected returns*.

18.2 ▪ Model

18.2.1 ▪ Prior Distribution

A first attempt to solve the first problem (mismatch between the equilibrium and the tangency portfolio) is to recognize that the true expected return is somewhere "in between" the equilibrium expected return (given by the market) and the investor's beliefs, but we do not know exactly where. The expected return $\tilde{\boldsymbol{\mu}}$ is modeled as a random variable. Since $\tilde{\boldsymbol{\mu}}(\omega)$ is a random variable, it must have a variance-covariance matrix, which we name $\tau\boldsymbol{\Sigma}$.[54]

Definition 18.2. *The equilibrium expected return $\boldsymbol{\pi}$ is the market expected return implied by mean-variance allocation, i.e.,*[55]

$$\boldsymbol{\pi} = r_f\mathbf{1} + (B - Ar_f)\boldsymbol{\Sigma}\mathbf{w}^M$$

The (nonrandom!) equilibrium expected return $\boldsymbol{\pi}$ is thus a good choice for the mean of the prior distribution for $\tilde{\mu}(\omega)$. We write $f_{\tilde{\mu}}$ for this **prior density**:

$$f_{\tilde{\boldsymbol{\mu}}}(\boldsymbol{\mu}) \sim \exp\left(-\frac{1}{2}(\boldsymbol{\mu} - \boldsymbol{\pi})^t(\tau\boldsymbol{\Sigma})^{-1}(\boldsymbol{\mu} - \boldsymbol{\pi})\right)$$

where the symbol $\sim$ means that we omit the normalization factor.

18.2.2 ▪ Modeling of the Views

One defines three objects:

- a pick matrix $\mathbf{P}$, which selects particular assets on which views are taken
- a view on the expected return $\tilde{\boldsymbol{\mu}}$, which we denote by $\mathbf{v}$
- uncertainty on the view, represented by the mean-zero random variable $\tilde{\mathbf{z}}$.

The BL model can thus be stated as

$$\boldsymbol{P}\tilde{\boldsymbol{\mu}}(\omega) = \tilde{\mathbf{v}}(\omega) + \tilde{\mathbf{z}}(\omega) \tag{18.2}$$

where $\tilde{\mathbf{z}}$ is a mean-zero normal random variable with variance-covariance matrix $\boldsymbol{\Omega}$, independent from $\tilde{\mathbf{u}}$.

A conceptual difficulty in understanding this model is that the view is (ex ante) a random variable $\tilde{\mathbf{v}}$ but (ex post) a nonrandom vector $\mathbf{v}$. We will always use in our calculations an "ex-post" value $\mathbf{v}$ for the view. Bayesian analysis mixes the information on the prior distribution on $\tilde{\boldsymbol{\mu}}$ and the uncertainty on the view by assuming that, before the analyst makes a decision, the view is ("ex-ante") a random variable.[56]

Example 18.3. Mean-variance analysis is the special case where we assume complete knowledge of the expected return $\tilde{\boldsymbol{\mu}}$, which is not random. If the CAPM works,

$$\tilde{\boldsymbol{\mu}} = \boldsymbol{\pi}$$

The pick matrix is the identity matrix, and the value of the view $\mathbf{v}$ is equal to the equilibrium expected return $\boldsymbol{\pi}$. Moreover, since the expected return is not random, in this example we have

$$\bar{\mathbf{r}} = \tilde{\boldsymbol{\mu}}$$

[54] It is simpler to set the imprecision matrix $\tau\boldsymbol{\Sigma}$ to be a multiple τ of the matrix $\boldsymbol{\Sigma}$, but it does not need to be.

[55] This is obtained by applying (16.14).

[56] This enables us to use Bayes's law.

Example 18.4. We believe that Germany will do 5% better than France on average. Suppose that there is no uncertainty on this view, i.e., $z(\omega) = 0$. If $\tilde{\mu}_1$ is Germany's expected return, and $\tilde{\mu}_2$ is France's expected return, the investor implements the relation

$$\tilde{\mu}_1(\omega) - \tilde{\mu}_2(\omega) = \tilde{v}(\omega)$$

In vector notation, we have

$$\mathbf{P}\tilde{\boldsymbol{\mu}}(\omega) = \tilde{v}(\omega)$$

where

$$\mathbf{P} = \begin{bmatrix} 1 & -1 \end{bmatrix}$$
$$v = 5\%$$

Since the pick matrix is rectangular, $\tilde{\mu}$ can still be random, even if the view is known (i.e., if $v(\omega) = v$).[57]

Example 18.5. One could have two views. Let

$$\tilde{\boldsymbol{\mu}} = \begin{bmatrix} \text{Germany's exp. return} \\ \text{France's exp. return} \\ \text{Spain's exp. return} \end{bmatrix}$$

We keep the first view from the previous example and add a second view: Spain will do worse than France by 2%±1%. This is translated into

$$\mathbf{P} = \begin{bmatrix} 1 & -1 & 0 \\ 0 & 1 & -1 \end{bmatrix}$$
$$\mathbf{v} = \begin{bmatrix} 5\% \\ 2\% \end{bmatrix}$$
$$\tilde{\mathbf{z}}(\omega) = \begin{bmatrix} 0 \\ \tilde{z}_2(\omega) \end{bmatrix} \qquad \boldsymbol{\Omega} = \begin{bmatrix} 0 & 0 \\ 0 & (1\%)^2 \end{bmatrix}$$

where $\tilde{z}_2(\omega)$ is a normal random variable with mean 0 and standard deviation 1%.

The conditional density $f_{\mathbf{v}|\tilde{\mu}}$ of the view given the expected return is then

$$f_{\tilde{\mathbf{v}}|\tilde{\boldsymbol{\mu}}}(\mathbf{v}|\boldsymbol{\mu}) \sim \exp\left(-\frac{1}{2}(\mathbf{v} - \mathbf{P}\boldsymbol{\mu})^t \boldsymbol{\Omega}^{-1}(\mathbf{v} - \mathbf{P}\boldsymbol{\mu})\right) \tag{18.3}$$

Now, the problem is to "invert" this relation. Given the particular view, of the mean, what should be the return? In other words, what is the conditional density (called **posterior density**) $f_{\tilde{\boldsymbol{\mu}}|\tilde{\mathbf{v}}}(\boldsymbol{\mu}|\mathbf{v})$ of the expected return mean $\tilde{\boldsymbol{\mu}}$ given the view, and given the prior?

Proposition 18.6.

$$f_{\tilde{\boldsymbol{\mu}}|\tilde{\mathbf{v}}}(\boldsymbol{\mu}|\mathbf{v}) \sim \exp\left(-\frac{1}{2}(\boldsymbol{\mu} - \boldsymbol{\mu}_{BL}(\mathbf{v}))^t \boldsymbol{\Sigma}_{BL}^{-1}(\boldsymbol{\mu} - \boldsymbol{\mu}_{BL}(\mathbf{v}))\right)$$

[57] Take a random variable $x(\omega)$, and set $\mu_1(\omega) = v + x(\omega)$ and $\mu_2(\omega) = x(\omega)$.

where

$$\begin{aligned}\boldsymbol{\Sigma}_{BL} &= ((\tau\boldsymbol{\Sigma})^{-1} + \mathbf{P}^t\boldsymbol{\Omega}^{-1}\mathbf{P})^{-1} \\ \boldsymbol{\mu}_{BL}(\mathbf{v}) &= \boldsymbol{\Sigma}_{BL}((\tau\boldsymbol{\Sigma})^{-1}\boldsymbol{\pi} + \mathbf{P}^t\boldsymbol{\Omega}^{-1}\mathbf{v})\end{aligned} \tag{18.4}$$

Proof. By Bayes's law,

$$f_{\tilde{\boldsymbol{\mu}}|\tilde{\mathbf{v}}}(\boldsymbol{\mu}|\mathbf{v}) = \frac{f_{\tilde{\mathbf{v}}|\tilde{\boldsymbol{\mu}}}(\mathbf{v}|\boldsymbol{\mu})}{f_{\tilde{\mathbf{v}}}(\mathbf{v})} f_{\tilde{\boldsymbol{\mu}}}(\boldsymbol{\mu})$$

You will see the remainder of the proof in Exercise 18.2. ■

We now return to the original problem. To summarize, we use two random variables to represent two types of risk:

- $\tilde{\mathbf{u}}(\omega)$ represents **market risk**
- $\tilde{\mathbf{z}}(\omega)$ represents **ambiguity risk**

and we determine the conditional distribution of the return, given the view.

Proposition 18.7.

$$f_{\tilde{\mathbf{r}}|\tilde{\mathbf{v}}}(\mathbf{r}|\mathbf{v}) \sim \exp\left(-\frac{1}{2}(\mathbf{r} - \boldsymbol{\mu}_{BL}(\mathbf{v}))^t(\boldsymbol{\Sigma} + \boldsymbol{\Sigma}_{BL})^{-1}(\mathbf{r} - \boldsymbol{\mu}_{BL}(\mathbf{v}))\right) \tag{18.5}$$

You will prove this proposition in Exercise 18.3. Observe that model risk (represented by $\boldsymbol{\Sigma}_{BL}$) adds to the variance of the returns. Indeed, the variance-covariance matrix grows from $\boldsymbol{\Sigma}$ to $\boldsymbol{\Sigma} + \boldsymbol{\Sigma}_{BL}$. Once these parameters are determined, the optimization problem for a given expected return μ becomes

$$\begin{aligned}&\min_{\mathbf{w}} \mathbf{w}^t(\boldsymbol{\Sigma} + \boldsymbol{\Sigma}_{BL})\mathbf{w} \\ &\mathbf{w}^t\boldsymbol{\mu}_{BL}(\mathbf{v}) = \mu\end{aligned}$$

Figures 18.1 and 18.2 illustrate the BL analysis that Krystal and Yimen Hong made in 2017.

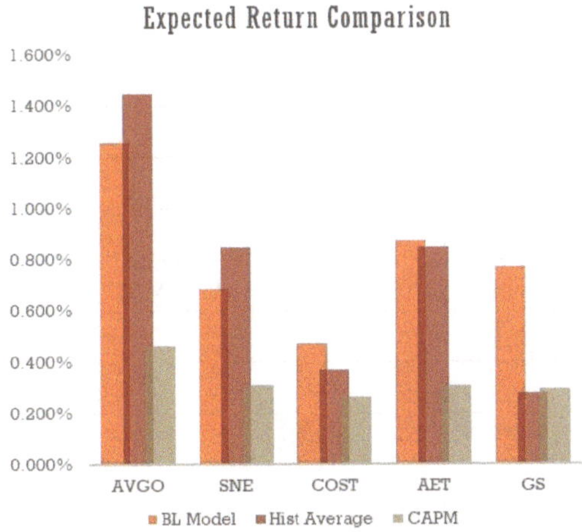

Figure 18.1: Investing in five companies using the Black–Litterman approach.

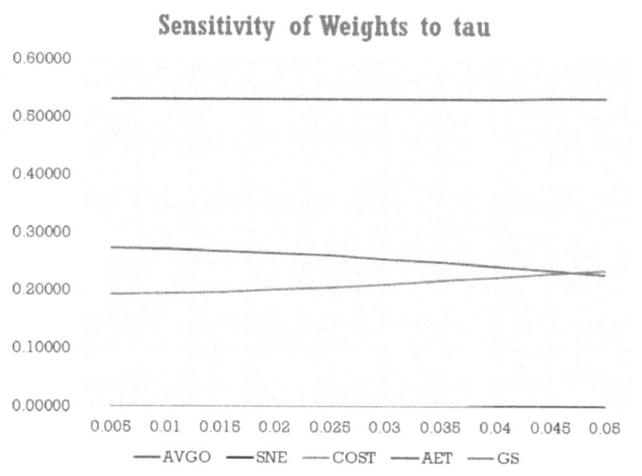

Figure 18.2: Black–Litterman: sensitivity of the weights to model risk.

18.3 ▪ Interpretation

In Proposition 18.7, we calculated the BL mean $\boldsymbol{\mu}_{BL}$ and variance $\boldsymbol{\Sigma}_{BL}$ directly from the densities. There is an alternative representation, which gives a much better intuition on these parameters.

Proposition 18.8.

$$\boldsymbol{\mu}_{BL} = \boldsymbol{\pi} + \tau\boldsymbol{\Sigma}\mathbf{P}('\mathbf{P}\tau\boldsymbol{\Sigma}\mathbf{P}'+\boldsymbol{\Omega})^{-1}(\mathbf{v} - \boldsymbol{P\pi}) \tag{18.6}$$

$$\boldsymbol{\Sigma}_{BL} = \tau\boldsymbol{\Sigma}-\tau\boldsymbol{\Sigma}\mathbf{P}'(\mathbf{P}\tau\boldsymbol{\Sigma}\mathbf{P}'+\boldsymbol{\Omega})^{-1}\mathbf{P}\tau\boldsymbol{\Sigma} \tag{18.7}$$

Proof. We prove only the result for $\boldsymbol{\Sigma}_{BL}$ in the case $\mathbf{P} = \mathbf{I}$, and take $\tau = 1$. The general case will be proved in the problem section. For notational simplicity, we write the right-hand side of (18.7) as $\mathbf{A}$. Thus

$$\mathbf{A} = \boldsymbol{\Sigma} - \boldsymbol{\Sigma}(\boldsymbol{\Sigma} + \boldsymbol{\Omega})^{-1}\boldsymbol{\Sigma}$$

while from (18.4)

$$\boldsymbol{\Sigma}_{BL}^{-1} = \boldsymbol{\Sigma}^{-1} + \boldsymbol{\Omega}^{-1}$$

We multiply both matrices with the hope that this gives us the identity matrix, and we obtain

$$\mathbf{A}\boldsymbol{\Sigma}_{BL}^{-1} = \mathbf{I} + \boldsymbol{\Sigma}\left[\boldsymbol{\Omega}^{-1} - (\boldsymbol{\Sigma} + \boldsymbol{\Omega})^{-1} - (\boldsymbol{\Sigma} + \boldsymbol{\Omega})^{-1}\boldsymbol{\Sigma}\boldsymbol{\Omega}^{-1}\right] \tag{18.8}$$

Elementary manipulations (start with $(\boldsymbol{\Sigma} + \boldsymbol{\Omega})\boldsymbol{\Omega}^{-1} = \boldsymbol{\Sigma}\boldsymbol{\Omega}^{-1} + \mathbf{I}$ and premultiply by $(\boldsymbol{\Sigma} + \boldsymbol{\Omega})^{-1}$) show that

$$\boldsymbol{\Omega}^{-1}=(\boldsymbol{\Sigma} + \boldsymbol{\Omega})^{-1}\boldsymbol{\Sigma}\boldsymbol{\Omega}^{-1} + (\boldsymbol{\Sigma} + \boldsymbol{\Omega})^{-1} \tag{18.9}$$

Inserting (18.9) in (18.8), we conclude that $\mathbf{A}\boldsymbol{\Sigma}_{BL}^{-1}$ is the identity matrix, as claimed. ■

The interpretation is easier when we take the view matrix to be the identity matrix. We consider two polar cases: infinite precision of the view and infinite variance of the view. Recall that *precision* is defined to be the inverse of the variance.

Fact 18.9. *When the view has an infinite variance, the (Black–Litterman) expected return is equal to the prior, and the (Black–Litterman) variance-covariance matrix is equal to the variance-covariance matrix of the prior, i.e.,* $\tau\Sigma$.

Fact 18.10. *When the view has infinite precision, the (Black–Litterman) expected return is equal to the view, and the (Black–Litterman) variance-covariance matrix is equal to zero.*

Exercises

18.1. (Hard) Prove relation (18.1) in the scalar case.

18.2. (Medium) Complete the proof of Proposition 18.6.

Hint: Calculate the terms that are quadratic and linear in $\boldsymbol{\mu}$ of -2 times the log of the right-hand side. You should obtain for the quadratic terms

$$\boldsymbol{\mu}'(\mathbf{P}^t\boldsymbol{\Omega}^{-1}\mathbf{P}+(\tau\boldsymbol{\Sigma})^{-1})\boldsymbol{\mu}$$

and for the linear terms

$$-2\boldsymbol{\mu}'(\mathbf{P}^t\boldsymbol{\Omega}^{-1}\mathbf{v}+(\tau\boldsymbol{\Sigma})^{-1}\boldsymbol{\pi})$$

Knowledge that $\tilde{\boldsymbol{\mu}}$ is conditionally normal completes the proof. (Why?)

18.3. (Medium) Prove Proposition 18.7.

18.4. (Hard) Complete the proof of Proposition 18.8 in the general case.

18.5. (Medium) Prove Facts 18.9 and 18.10. Convince yourself by developing a small 2-by-2 example.

18.6. We know that the expected return of the world portfolio is 4%, i.e.,

$$\mu - r_f = 0.04$$

Take a three-continent world (Europe, Asia, America). Assume that the variance-covariance matrix of returns is

$$\Sigma = \begin{bmatrix} 0.2 & 0.15 & 0.1 \\ 0.15 & 0.25 & 0.1 \\ 0.1 & 0.1 & 0.1 \end{bmatrix}$$

There is model risk around these parameters, represented by the prior variance-covariance matrix

$$\tau\Sigma = \tau * \Sigma$$

where $\tau = 1/2$. Assume that each continent's capitalization is equal to a third of the world's capitalization and that the risk-free rate is 2%. You want an expected return of 3% on your portfolio. In your opinion, Europe is going to do better than Asia by $1\%\pm1\%$. What should your portfolio be?

18.7. In this problem, you will construct a world portfolio. Use real data to calculate the historical return of three main global assets: US stock, Euro stock, and Chinese stock. By "stock," we mean the total capitalization of the stock (all stocks in each country, or a proxy thereof, such as S&P 500). Feel free to choose any observation period.
Then calculate the observed mean return $\bar{\mathbf{r}}^{obs}$ vector and covariance matrix of return $\mathbf{\Sigma}^{obs}$. Assume that your estimation is correct, i.e., the true mean return $\bar{\mathbf{r}}$ is equal to $\bar{\mathbf{r}}^{obs}$, and the true covariance matrix of return $\mathbf{\Sigma}$ is equal to $\mathbf{\Sigma}^{obs}$.

(a) Calculate the "market portfolio" $\mathbf{w}^M$, as well as the world expected return $E[\mathbf{r}^M]$. Here we assume that these three countries represent all the world economy.

(b) Is the world portfolio equal to the tangent portfolio? In other words, verify how close $\mathbf{w}^M$ is to $\mathbf{w}^T$.

You should provide numerical values for $\bar{\mathbf{r}}, \mathbf{\Sigma}, A, B, C, r_f, \mathbf{w}^M, \mathbf{w}^T$, and $E[\mathbf{r}^M]$.
As we do not provide data for this question, specify which series you used for US stock, Euro stock, and Chinese stock. Feel free to choose any period you want.

18.8. **Project:** Apply the Black–Litterman methodology to the data of your choice. Do not use a package. Code the solution from scratch in either Excel or MATLAB.

Chapter 19

Strategic Asset Management

In this chapter, we generalize the static portfolio problem seen in Chapter 14 to a dynamic problem. In the **dynamic portfolio choice** problem,[58] the decision-maker has the opportunity to adjust the portfolio weights at several intermediate times between now, that is, time 0, and the horizon T. It does not mean that the resulting optimal plan is rigid. After choosing her portfolio today, the investor can and should change her portfolio tomorrow on the basis of new information. However, considering long-term changes in prices brings a different ("strategic") solution from the static problem. The difference is called the *hedging demand.* It takes into account the fact that bond prices, as well as volatility, are stochastic and sometimes mean-reverting. It can incorporate mean-reversion in commodity prices. Because of mean-reversion, the strategic allocation tends to be more contrarian than the tactical one.

So far this book has focused only on asset management. In practice, banks and insurance companies must manage at the same time the *asset* and *liability* sides of their portfolio. We assume in this chapter that the liabilities are given. We will consider two different types of markets: competitive and noncompetitive markets. In competitive markets, an order to buy/sell a security is executed at the prevailing market price, regardless of the quantity of the order. In the other situation there are transaction costs and illiquidity effects: the larger a buy order, the higher the transaction price will be.

19.1 ▪ Competitive Markets

We will cover the "father" of all parametric strategic asset management models, the Merton model, and then an ML version, following the martingale approach. But before that, we develop some accounting relations of self-financing portfolios.

19.1.1 ▪ Self-Financing Portfolios

Trading occurs at integer multiples of the time step Δt, that is,

$$t \in \{0, \Delta t, 2\Delta t, \ldots, T = N\Delta t\}$$

Fix t. The value of the portfolio just before rebalancing at a time $t + \Delta t$ is thus

$$W^c(t + \Delta t, \omega) = \underbrace{\sum_{i=0}^{I} n_{i,t}(\omega) x_{i,t+\Delta t}(\omega) - \underbrace{L_{t+\Delta t}(\omega)}_{\text{liability payment}}}_{\text{portfolio value}}$$

[58] Also referred to as either "strategic asset management" or "strategic asset allocation."

where $n_{i,t}$ is the number of shares of security i held at time t, and $x_{i,t+\Delta t}(\omega)$ is the payoff at time $t + \Delta t$.

$$\underbrace{x_{i,t+\Delta t}(\omega)}_{\text{payoff}} = \underbrace{d_{i,t+\Delta t}(\omega)}_{\text{dividend}} + \underbrace{p_{i,t+\Delta t}(\omega)}_{\text{price ex-dividend}}$$

Just after receiving the dividends, at time $t + \Delta t$ the asset manager optimally rebalances his portfolio; that is, she chooses a new number of shares of each security, $n_{i,t+\Delta t}(\omega)$, resulting in wealth,

$$W(t + \Delta t, \omega) = \sum_{i=0}^{I} n_{i,t+\Delta t}(\omega) p_{i,t+\Delta t}(\omega)$$

The portfolio is *self-financing*; that is, the wealth before rebalancing and after rebalancing must be the same:

$$W^c(t + \Delta t, \omega) = W(t + \Delta t, \omega) \tag{19.1}$$

19.1.2 ▪ The Merton Problem

The goal of the portfolio manager is to maximize the expected utility of wealth at the horizon:

$$\max E[u(W(T))]$$

As in Chapter 17, we suppose that investors are risk-averse, that is, u is concave, increasing, and twice differentiable. We now look at a simplified version of the problem with only one risk-free asset, asset 0, with a deterministic return equal to r_f, and one risky asset, asset 1. We can get rid of the subscript $_{i=1}$ for the risky asset and write its return as

$$r(t, t + \Delta t, \omega) = \frac{p(t + \Delta t, \omega) - p(t)}{p(t)}$$

As shown in Chapter 16, the return of the portfolio is

$$r_p(t, t + \Delta t, \omega) = (1 - w(t))r_f + w(t)r(t, t + \Delta t, \omega)$$

where $w(t)$ is the amount of money invested in the risky asset. Moreover, we assume that the return is independent of the level of the stock price. Our model is arithmetic **Brownian motion**:[59]

$$r(t, t + \Delta t, \omega)\Delta t = \bar{r}\Delta t + \sigma z(t + \Delta t, \omega)\sqrt{\Delta t} \tag{19.2}$$

where $z(t + \Delta t)$ is a standard normal random variable, realized at time $t + \Delta t$. The variables z at all times are supposed IID. We have then

$$E_t[r(t, t + \Delta t)\Delta t] = \bar{r}\Delta t$$
$$Var_t[r(t, t + \Delta t)\Delta] = \sigma^2 \Delta t$$

The quantities $\bar{r}$ and σ are called, respectively, the **drift** and **volatility** of arithmetic Brownian motion. Moreover, if Δt is small, we can make the approximation

$$E_t[(r(t, t + \Delta t)\Delta t)^2] \simeq Var_t[r(t, t + \Delta t)\Delta t]$$

[59](Standard) Brownian motion is the special case of arithmetic Brownian motion with no drift and unit volatility. Another name for Brownian motion is the Wiener process.

By definition of the portfolio return and the equality of wealth before and after rebalancing (19.1),

$$W(t+\Delta t,\omega) = (1 + r_p(t, t+\Delta t,\omega)\Delta t)W(t,\omega) - L(t,\omega)$$

It is now a good idea to define the following stochastic process.[60]

Definition 19.1. $\Delta W(t) = W(t+\Delta t) - W(t)$.

For ease of exposition, we suppose now that the liability $L(t,\omega) = 0$. Inserting our model for the return (19.2), and getting rid of the ω argument to lighten up the notation, we obtain the **state equation**

$$\Delta W(t) = W(t)w(t)(\bar{r} - r_f)\Delta t + \sigma w(t)z(t+\Delta t)\sqrt{\Delta t}$$

where wealth $W(t)$ is the state and the risky asset weight $w(t)$ is the control. We solve the problem by dynamic programming. The value function is now

$$J(t, W(t)) = \max_{\text{all controls}} E[u(W(T))|W(t)]$$

As shown in Chapter 14, the value function must solve the HJB equation, but for the Mayer problem

$$J(t, W(t)) = \max_{w(t)} E[J(t+\Delta t, W(t+\Delta t))|W(t)] \tag{19.3}$$

To obtain an analytical solution, the best approach is calculus. Using a Taylor approximation,

$$J(t+\Delta t, W(t)+\Delta W(t)) \simeq J(t, W(t)) + \frac{\partial J}{\partial t}\Delta t + \frac{\partial J}{\partial W}\Delta W(t) + \frac{1}{2}\frac{\partial J^2}{\partial W^2}[\Delta W(t)]^2$$

The term $J(t, W(t))$ cancels out in the HJB equation (19.3), which becomes

$$0 \simeq max_{w(t)} E[\frac{\partial J}{\partial t}\Delta t + \frac{\partial J}{\partial W}\Delta W(t) + \frac{1}{2}\frac{\partial J^2}{\partial W^2}(\Delta W(t))^2|W(t)] \tag{19.4}$$

As you will see in the exercise section, the conditional first moment of $\Delta W(t)$ per unit of time is $W(t)w(t)[\bar{r} - r_f]$. Its conditional second moment, also per unit of time, is $(W(t)w(t)\sigma)^2$. We can thus divide (19.4) by Δt to obtain

$$0 = \max_{w(t)} \left(\frac{\partial J}{\partial t} + \frac{\partial J}{\partial W} W(t)w(t)(\bar{r} - r_f) + \frac{1}{2}\frac{\partial J^2}{\partial W^2} W(t)^2 w(t)^2 \sigma^2 \right)$$

A Particular Solution

Suppose $u(W(T)) = \ln(W(T))$. We now look for a separable value function and try the solution $J(t, W(t)) = k(t)\ln(W(t))$ for some function $k(t)$. Inserting this in the (HJB) equation, we obtain

$$0 = \max_{w(t)} \left(\frac{\partial J}{\partial t} + k(t)\left(w(t)[\bar{r} - r_f] - \frac{1}{2} w(t)^2 \sigma^2 \right) \right)$$

Differentiating with respect to the weight, we obtain the optimal allocation:

$$w^*(t) = \frac{\bar{r} - r_f}{\sigma^2} \tag{19.5}$$

This allocation rule is also known as the **Kelly criterion**.

[60] It is the only stochastic process in this book that is not adapted, i.e., not known at time t.

Generalizations

Let us return now to the general problem with I securities. As in Chapter 16, the mean return vector is denoted by $\bar{\mathbf{r}}$ and the variance-covariance matrix by $\boldsymbol{\Sigma}$. Suppose that the utility function is of power type (defined in the exercise). You will show in the exercise the following result, which is the multivariate equivalent of (19.5).

Theorem 19.2 (Merton). *The optimal proportion of wealth allocated to each stock is for logarithm and power utility functions*

$$\mathbf{w}^*(t) = \gamma \boldsymbol{\Sigma}^{-1}(\bar{\mathbf{r}} - r_f \mathbf{1}) \tag{19.6}$$

where $1/\gamma$ is the degree of risk aversion of the investor.

Observations. This is **exactly** the Markowitz solution (Theorem 16.2) from static portfolio choice. As important is the fact that the optimal weight is **constant**. Recall, however, that a constant portfolio weight $\mathbf{w}^*$ does not mean that the number of stocks held $\mathbf{n}^*$ should be constant. At each period, the asset manager should rebalance her portfolio: if a stock price goes up, she should sell that stock. If it goes down, she should buy it. This is a contrarian result.

One could adopt a more complicated return model than arithmetic Brownian motion: adding stochastic volatility and/or mean-reverting returns or interest rates. In that case, the optimal weight $\mathbf{w}^*(t)$ is not constant anymore. A popular term for it is a **glide path** solution. The optimal allocation turns out to be even more contrarian than in the arithmetic Brownian motion case.

19.1.3 ▪ A Machine Learning Model

Let us anticipate on the notation of Chapter 26. The information available to market participants at time t is denoted by $\mathbf{X}_t \equiv \{\mathbf{x}_{1,t}, \dots, \mathbf{x}_{I,t}\}$. It includes market-specific information, as well as idiosyncratic information. At time zero, the asset manager invests in a portfolio, knowing that she can rebalance it optimally until her horizon, time T. The economics department[61] of her firm forecasts some scenarios $\mathbf{X}_t(\omega)$ of the state variables, with $\omega = 1, \dots, \Omega$, with $t = 1, \dots, T$.

It also includes forecasts of some security prices $\mathbf{p}_t(\omega)$. For simplicity, we assume that each scenario has the same probability $\frac{1}{\Omega}$.

With this information in hand, the asset manager can forecast the stochastic discount factor (SDF) $m(t, \omega)$, as we will see in more detail in Chapter 25. By no-arbitrage (see relation (22.3)), the portfolio wealth at time zero is equal to the expected value of the product of the SDF and wealth at the horizon T:

$$W(0) = E[m(T)W(T)]$$

We replace expectations by averages and represent the portfolio problem as

$$\max \frac{1}{\Omega} \sum_{\omega=1}^{\Omega} u(W(T, \omega))$$

$$\frac{1}{\Omega} \sum_{\omega=1}^{\Omega} m(T, \omega) W(T, \omega) = W(0) \tag{19.7}$$

We deliberately omitted the decision variables in the formulation of this problem. We are indeed going to take a two-step approach to this problem:

- first solve for the optimal wealth at the horizon $W^*(T, \omega)$
- then calculate the portfolio $\mathbf{n}(t, \omega)$ that results in the optimal wealth $W^*(T, \omega)$.

[61] In the US, the Federal Reserve delivers quarterly $\boldsymbol{\Omega} = 16$ scenarios of economic forecasts.

This is the essence of the **martingale approach**, pioneered by Cox and Huang [62] and Karatzas, Lehoczy, and Shreve [63]. The optimal portfolio can be calculated by the Lagrangian approach.

First Problem: Determining the Optimal Wealth This is a classical programming problem with Ω decision variables $W^*(T,\omega)$. Calling λ the Lagrange multiplier of the constraint (19.7), the first-order conditions are

$$u'(W^*(T,\omega)) = \lambda m(T,\omega) \quad \omega = 1,\ldots,\Omega \tag{19.8}$$

We invert this relation for the optimal wealth and then determine λ by inserting $W^*(T,\omega)$ back in the constraint (19.7). For instance, when the utility function is logarithmic, we obtain

$$W^*(T,\omega) = \frac{W_0}{m(T,\omega)}$$

Otherwise, the Lagrange multiplier can be obtained by a one-dimensional search.

Second Problem: Determining the Optimal Number of Shares For simplicity, we assume no dividends. Then the wealth at time t can be calculated by evaluating two different scalar products: price at time t times number of shares of time t (after rebalancing), or price at time t times number of shares of time $t-1$ (before rebalancing). The first idea underlying the following algorithm is to use the former representation. The second idea is to realize that $m(t)W(t)$ is a martingale in the physical measure. Since every martingale is a conditional expectation, and since every conditional expectation solves the least squares problem (see Algorithm 14.3), we obtain the optimal portfolio by regression.

For the logarithmic utility function, the algorithm is deceptively simple.

ALGORITHM 19.1. Nonparametric martingale algorithm.

For $\omega = 1,\ldots,\Omega$
    Simulate $p(1,\omega)$ and $m(1,\omega)$
Regress $W_0/m(1,\omega)$ on $[\ 1 \quad \mathbf{p}(1,\omega)\ ]$
Denote the coefficients of the regression by a and $\mathbf{b}$
Set $\mathbf{n}(0) = \mathbf{b}$

Observations. We could generalize this algorithm to determine the optimal dynamic allocation $\mathbf{n}^*(t,\omega)$ for all times t between time zero and the horizon by regressing wealth over future SDF on future prices. However, in practice, portfolio managers will receive at each time fresh information, i.e., a new forecast of the economy and of the prices. As an example, let us run the plan at time 0, which will give us an optimal number of shares $\mathbf{n}^*(1,\omega)$ at time one. Time passes, and today is now time one. If the investor has fresh information at time one, he should run the algorithm again to obtain a new optimal number of shares $\mathbf{n}^*(\mathbf{1})$ which is potentially different from the previously obtained solution $\mathbf{n}^*(1,\omega)$. If the forecast has not changed, then the solutions will be identical.

It is thus a very economical algorithm. It, however, makes sense only if there are more scenarios than securities.

This algorithm considers only the logarithm utility function. It can easily be generalized to power utility functions. In that case, a one-dimensional search for the parameter λ in (19.8) is needed. See Shreve [64] for more details. For more complicated utilities than the (myopic) logarithmic utility function, it will become necessary to forecast the SDF until the horizon T.

In that case, the strategic advantage of this algorithm becomes clear: not only short-term but also long-term forecasts of the SDF will influence the optimal portfolio.

19.2 ▪ Noncompetitive Markets

We now discuss transaction costs and illiquidity effects. Transaction costs will not change the quoted price of a security. Illiquidity effects will: there is a price impact of a large order on the market. For the modeling of transaction costs in a linear program, we refer the reader to Chapter 5.

There is a large literature on illiquidity effects. In this case, there are strategies for optimal trade execution. The main idea is to split the order in smaller pieces, with the hope that "prices will return to normal," that is, the impact of a large order is not permanent. There will be a tradeoff between the loss incurred by moving the market (by placing a large order) and the risk of waiting (incurred when submitting a large quantity of successive small orders).

Gârleanu and Pedersen [67] consider the general problem of **optimal stock execution** when trading multiple securities. They incorporate in the Merton portfolio problem illiquidity costs. They use a quadratic utility function, where the utility is equal to the expected return minus a risk-aversion coefficient γ times the variance minus the trading costs.

Definition 19.3. *The speed of trading $\tau_{i,t}$ of trading security i is equal to $n_{i,t} - n_{i,t-1}$.*

Trading $\boldsymbol{\tau}_t$ shares will result in a quadratic cost if we suppose that the price adjusts linearly to a large trade: $p_i(t) \to p_i(t) + \lambda\tau_{i,t}$. The illiquidity cost of buying $\tau_{i,t}$ shares is thus equal to $\lambda\tau_{i,t}^2$, where $\lambda \geq 0$ is some measure of illiquidity. We generalize this to an illiquidity matrix $\boldsymbol{\Lambda}$, supposed positive semidefinite.

Definition 19.4. *The total illiquidity cost $TC(T)$ up to time T is modeled as $E[\sum_{t=1}^{T} \boldsymbol{\tau}_t\boldsymbol{\Lambda}\boldsymbol{\tau}_t]$.*

The problem is then to

$$\max_{\mathbf{n}} E[W(T)] - \gamma Var[W(T)] - TC(T) \tag{19.9}$$

where $\mathbf{n}_t$ is, as before, a nonanticipative control: it can be calculated based only on information at time t. The net effect of the illiquidity cost is to smooth the control, as shown in Figure 19.1. The optimal weights tend to follow a long-term aimed portfolio. The regularizing effect of transaction costs is not surprising: we discussed the importance of regularization elsewhere in the book.

19.3 ▪ To Probe Further

This material is much easier to follow in continuous time if you know some basic stochastic calculus. We recommend reading the books by Duffie [68], Korn [18], or Pham [46]. There is a large literature on strategic asset allocation, especially in academia, where this area is known as dynamic portfolio choice. Almgren and Chriss [65] consider two types of impact of large orders on price: permanent and transient. This type of concern is important for high-frequency traders. A survey of techniques used in high-frequency trading is given in Cartea, Jaimungal, and Penalva [66].

The nonparametric martingale algorithm is, to the best of our knowledge, an original algorithm. We have not tested it thoroughly and do not advise asset managers to use it until conscientious quants study it more closely.

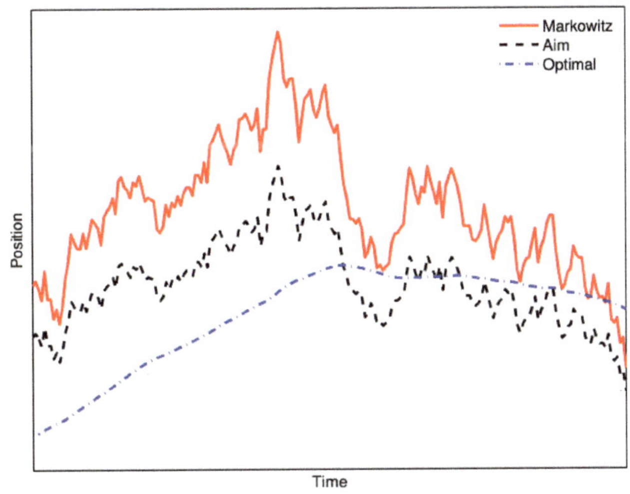

Figure 19.1: Optimal portfolio with one asset and temporary impact costs. From Gârleanu and Pedersen [67] under a CC BY 4.0 DEED license https://creativecommons.org/licenses/by/4.0/deed.en.

Exercises

19.1. (Medium) Prove that the conditional first moment of $\Delta W(t)$ per unit of time is $W(t)w(t)[\bar{r} - r_f]$. Prove that its conditional second moment, also per unit of time, is $(W(t)w(t)\sigma)^2$.

19.2. (Hard) Prove Theorem 19.2 for

(a) $u(W) = \ln(W)$

(b) power utility functions: $u(W) = W^\delta/\delta$ for $0 < \delta < 1$

19.3. (Hard) One is tempted in the Gârleanu–Pedersen model to solve it as a linear-quadratic problem, either numerically or analytically, using our solution from Exercise 3.7. For thus it would "suffice" to simulate the prices $\mathbf{p}(t,\omega)$ by Monte Carlo simulation and introduce the constraints

$$\begin{aligned} W(t,\omega) &= \mathbf{n}(t,\omega)'\mathbf{p}(t,\omega) \\ w_{i,t}(\omega) &= \frac{n_{i,t}(\omega)p_i(t,\omega)}{W(t,\omega)} \\ n_{i,t}(\omega) - n_{i,t-1}(\omega) &= \tau_{i,t}(\omega) \end{aligned}$$

Why is this a bad idea?

19.4. Implement the nonparametric martingale algorithm. Vary the utility functions. Try to implement dividends. To our knowledge, this idea has never been published. Try to submit the result to a journal, or, if you are a Ph.D. student, to convince your Ph.D. advisor that it could be a good idea for a dissertation.

Chapter 20
Capital Asset Pricing Model (CAPM)

We saw that the goal of mean-variance analysis is to decide, given statistics on securities returns, what the optimal investment should be for a single investor. The goal of a *pricing model* is to determine what the price statistics of each security should prevail in equilibrium, given that each investor optimizes his/her portfolio.

At first glance, it looks like the excess expected return of an investment in equilibrium should be inversely proportional to its volatility. In other words, every investment should have the same Sharpe ratio.

The CAPM shows that this is not true. Net expected returns should be proportional to the **covariance** between the security's return and the market return. Thus, investments with low or negative covariance with the market portfolio (such as gold or the Swiss franc) may have a lower Sharpe ratio than investments with a high covariance. Viewed a different way, in equilibrium, the expected return of gold can be lower than the expected return of other commodities that have the same volatility.

As stated earlier (but this point cannot be overemphasized), the CAPM is a one-factor model, where the single factor is the market portfolio excess return. Regressing the excess stock return of every security on the excess market return should yield no intercept. If there is an intercept, and this intercept is called **alpha**, then it is a sign that the security is mispriced. If alpha is positive, the security should be bought. Otherwise, it should be sold. This assumes of course that past returns are an indication of future returns.

20.1 ▪ Derivation

In this section, we derive the CAPM, using the original method proposed by Sharpe and others, as a natural extension of mean-variance analysis. We will see the more modern and equivalent (actually, more general) way to derive it in another section in Chapter 21.

In this version of the model, we assume that at each time t, each investor j maximizes a function $V_j(x, y)$, called the indirect utility function, such that

$$E_t[u_j(r^j_{t+1})] = V_j(E_t[r^j_{t+1}], Var_t[r^j_{t+1}]) \tag{20.1}$$

where u_j is the utility of investor j, and r^j_{t+1} is the net return of investor j's portfolio.

In other words, only the mean and variance of the portfolio return matter to determine an investor's preferences. As in the previous chapter, we will drop the subscript t, but it is important to remember that, like mean-variance, there is a *conditional* version of the CAPM, as well as an *unconditional* one. Clearly, investors prefer more expected return to less expected return, i.e.,

$$\frac{\partial V_j}{\partial x} > 0 \tag{20.2}$$

and less risk to more risk:

$$\frac{\partial V_j}{\partial y} < 0 \tag{20.3}$$

More precisely, the assumptions behind equilibrium in the CAPM are as follows:

1. each investor j maximizes a function V_j as above
2. all investors have a common time horizon and beliefs
3. each asset is infinitely divisible
4. the riskless asset can be bought or sold in unlimited amounts.

As usual, we drop the time subscript.

Definition 20.1. *The weight of security i in the market portfolio is defined by*

$$w_i^M - \frac{\textit{market cap of security } i}{\textit{total market cap}}$$

The net expected return on the market portfolio is denoted by $\bar{r}^M$.

We recall from Chapter 16 the definition of the **tangency portfolio**:

$$\mathbf{w}^T = \frac{\boldsymbol{\Sigma}^{-1}(\bar{\mathbf{r}} - r_f \mathbf{1})}{B - r_f A} \tag{20.4}$$

Theorem 20.2 (CAPM).

1. *In equilibrium, the tangency portfolio is the market portfolio, i.e.,*

$$\mathbf{w}^T = \mathbf{w}^M$$

2. *The excess return is given by*

$$\bar{\mathbf{r}} - r_f \mathbf{1} = \boldsymbol{\beta}^M (\bar{r}^M - r_f)$$

where the beta of security i is given by

$$\beta_i^M = \frac{Cov[r_i, \mathbf{w}^M \mathbf{r}]}{Var[\mathbf{w}^M \mathbf{r}]}$$

Proof. To avoid clutter, we drop the transpose subscript on $\mathbf{w}$. Let $\mathbf{w}^j$ be the wealth proportion of investor j. Thus

$$r^j = r^f + \mathbf{w}^j(\mathbf{r} - r_f \mathbf{1})$$

The optimization program of investor j is

$$\max_{\mathbf{w}^j} V_j(r^f + \mathbf{w}^j(\bar{\mathbf{r}} - r_f \mathbf{1}), \mathbf{w}^j \boldsymbol{\Sigma} \mathbf{w}^j)$$

The first-order conditions are, for each security i,

$$\frac{\partial V_j}{\partial w_i^j} = 0 \tag{20.5}$$

We obtain, for each security, i,

$$\begin{aligned}\frac{\partial V_j}{\partial w_i^j} &= \frac{\partial V_j}{\partial x}\frac{\partial[r^f + \mathbf{w}^j(\bar{\mathbf{r}} - r_f\mathbf{1})]}{\partial w_i^j} + \frac{\partial V_j}{\partial y}\frac{\partial \mathbf{w}^j\boldsymbol{\Sigma}\mathbf{w}^j}{\partial w_i^j}\\ &= \frac{\partial V_j}{\partial x}(\bar{r}_i - r_f) + 2\frac{\partial V_j}{\partial y}\sum_k \boldsymbol{\Sigma}_{ik}w_k^j\end{aligned}$$

The first-order conditions (20.5) are the following system of linear equations:

$$\frac{\partial V_j}{\partial x}(\bar{\mathbf{r}} - r_f\mathbf{1}) + 2\frac{\partial V_j}{\partial y}\boldsymbol{\Sigma}\mathbf{w}^j = 0$$

with solution

$$\mathbf{w}^j = c_j\mathbf{w}^T \tag{20.6}$$

where $\mathbf{w}^T$ is given by (20.4), and

$$c_j = \frac{1}{B - Ar_f}\frac{-\frac{\partial V_j}{\partial x}}{\frac{\partial V_j}{\partial y}}$$

Observe that the constant c_j is investor-specific. Thus, each investor j invests $c_j\mathbf{w}^T$ in stocks and $1 - c_j\mathbf{1}'\mathbf{w}^T$ in the risk-free asset. In other words, John invests a different proportion of his wealth in stocks than Mary does, but given this amount invested in stocks, he invests in each stock in the same proportion as Mary:

$$\frac{w_i^{John}}{\sum w_i^{John}} = \frac{w_i^{Mary}}{\sum w_i^{Mary}}$$

John and Mary (and every other investor) hold the same stock portfolio. In equilibrium, this portfolio must be the market portfolio; thus

$$\mathbf{w}^M = \mathbf{w}^T$$

The rest of the proof requires some algebra. Since the market return is defined as $r^M = \mathbf{w}^M\mathbf{r}$, we obtain

$$\begin{aligned}Var[\mathbf{w}^M\mathbf{r}] &= Var[\mathbf{w}^T\mathbf{r}]\\ &= \frac{C - 2r_fB + r_f^2A}{(B - r_fA)^2}\end{aligned} \tag{20.7}$$

and

$$E[\mathbf{w}^M\mathbf{r}] - r_f = \frac{C - 2r_fB + r_f^2A}{B - r_fA} = (B - r_fA)Var[\mathbf{w}^M\mathbf{r}] \tag{20.8}$$

Premultiplying (20.4) by $\boldsymbol{\Sigma}$, we obtain

$$\begin{aligned}\bar{\mathbf{r}} - r_f\mathbf{1} &= (B - r_fA)\boldsymbol{\Sigma}\mathbf{w}^M\\ &= (B - r_fA)Cov[\mathbf{r}, \mathbf{w}^M\mathbf{r}]\end{aligned} \tag{20.9}$$

From (20.7) and (20.8) we obtain

$$B - r_f A = \frac{E[\mathbf{w}^M \mathbf{r}] - r_f}{Var[\mathbf{w}^M \mathbf{r}]} \tag{20.10}$$

Thus, putting together (20.9) and (20.10), we find

$$\mathbf{r} - r_f = \frac{Cov[\mathbf{r}, \mathbf{w}^M \mathbf{r}]}{Var[\mathbf{w}^M \mathbf{r}]} (E[\mathbf{w}^M \mathbf{r}] - r_f)$$

■

Interpretation: In scalar form the CAPM states that the alpha α_i of a security should be zero, where

$$\bar{r}_i - r_f = \alpha_i + \beta_i^M (\bar{r}^M - r_f)$$

Thus, if one finds through regression that α_i is positive, one should buy that security. Conversely, one should sell security i if α_i is negative. This is why asset management is often called "a **quest for alpha**."

Figure 20.1 shows the isocontours of maximal indirect utility function V_1 and V_2. They are tangent to the global efficient frontier.

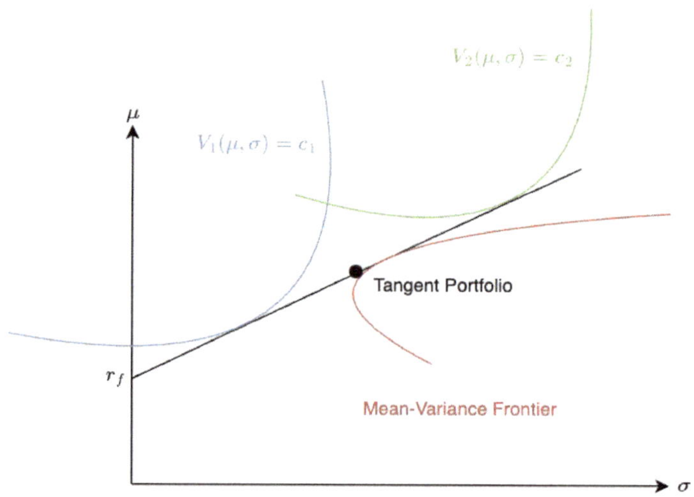

Figure 20.1: The blue investor is more risk-averse than the green investor. The blue investor will keep money in the bank. The green investor will borrow to buy stocks.

20.2 ▪ Consistency between Mean-Variance and Expected Utility Maximization

For what distributions of the return $\mathbf{r}$ and for what utility functions u does (20.1) hold with (20.2), (20.3)? In other words, for what return distributions and utility functions is the derived utility of wealth increasing in expected wealth and decreasing in wealth volatility?

It holds in two cases:

- quadratic utility function and any distribution
- any concave increasing utility function with an elliptical distribution (see section 20.2.2).

However, these results are not particularly limiting if one considers investing in continuous time. As mentioned in Chapter 16, an important result from probability[62] shows that a stochastic process $X(t)$ can have any distribution at a finite horizon $T > 0$ even if its conditional distribution[63] is normal. This is achieved by having the proper model for the *stochastic volatility*. Thus, **the conditional CAPM always holds** if one reinvests very frequently.

20.2.1 ▪ Quadratic Utility Function

The quadratic utility function is, for $b > 0$,

$$u(r) = r - \frac{b}{2}r^2$$

It is concave, but its main disadvantage is that for $r > 1/b$, the utility is decreasing. Thus, b needs to be calibrated well. You will show in the exercise section that mean-variance is consistent with expected utility maximization in this case.

20.2.2 ▪ Elliptical Distribution of Return

Elliptical distributions are distributions such that the isoprobability contours are ellipsoidal. A random vector $\boldsymbol{x}$ has an elliptical distribution if its density function is, for some function g and parameter I,

$$f(\boldsymbol{x}) = \frac{1}{|\boldsymbol{\Sigma}|^{1/2}} g((\mathbf{x} - \boldsymbol{\mu})'\boldsymbol{\Sigma}^{-1}(\mathbf{x} - \boldsymbol{\mu}); I)$$

where $\boldsymbol{\mu}$ is the vector of medians and $\boldsymbol{\Sigma}$ is proportional to the variance-covariance matrix. Normal and, more generally, t random variables are elliptically distributed. You will show in the exercise section that mean-variance is consistent with expected utility maximization in the simpler case of normal distributions.

Exercises

20.1. (Medium) For the quadratic utility function, give conditions such that the optimizer (20.6) of the expected utility maximization problem is indeed a maximizer and not a minimizer.

20.2. (Easy) Draw a capital market line and the contours of iso-derived utility in the quadratic utility case for some values of b.

20.3. (Medium) Consider the market model of Exercise 16.6. Can firms have different market capitalizations in this model?

20.4. (Medium) Show that $\frac{\partial V_j}{\partial x} > 0$ and $\frac{\partial V_j}{\partial y} < 0$ in the quadratic utility case.

20.5. (Hard) Show that $\frac{\partial V_j}{\partial x} > 0$ and $\frac{\partial V_j}{\partial y} < 0$ when returns are normally distributed, and the investor is risk-averse.

[62]The martingale representation theorem.

[63]For this, we need a short time frame, i.e., the distribution of $X(t + dt)$ given the information at time t has to be normal.

Chapter 21

Consumption-Based Asset Pricing

Whereas Chapters 16 (mean-variance) and 20 (CAPM) present the traditional, MBA-level explanation of asset pricing, this section presents a more academic view of the problem. It is, in some sense, more natural. In this chapter, the uncertainty is not represented solely by the mean and the variance of the asset, but by the complete probability distribution π of outcomes.

The CAPM views the portfolio problem as a pure problem of maximizing the expected utility of wealth. Lucas developed in the 1970s consumption-based asset pricing [69], which embeds the problem of wealth maximization in the larger problem of wealth/consumption tradeoff. Consumption-based asset pricing has the further advantage of providing a simple economic characterization of the stochastic discount factor. The main disadvantage is that relationship (21.3) does not hold in practice (see Exercise 21.3 on the equity premium puzzle) and is thus useful only for economists. Not for investors.

21.1 ▪ Basic Pricing Equation

In this chapter, we consider a world with a single risky asset. In the next chapter, we will generalize it to multiple assets. At each point in time, t, an investor has the choice between

- consuming now c_t
- buying a quantity n_t of the risky asset.[64]

The basic question is, what should be the price p_t at time t of the asset? We will use a basic one-period model, where one unit of the asset has a random payoff $x_{t+1}(\omega)$ at time $t+1$, which will be used for consumption at time $t+1$. The answer is that

loss in today's utility of consumption =
expected increase in (discounted) future utility of consumption

We denote by $\beta < 1$ the discount factor of future consumption, which reflects the fact that the investor prefers to consume now (t) rather than later ($t+1$).

The (intertemporal) utility function of the investor is then

$$U(c_t, c_{t+1}) = u(c_t) + \beta E_t[u(c_{t+1})]$$

[64] All variables c_t, n_t, p_t are known at time t, so we do not write $c_t(\omega), n_t(\omega), p_t(\omega)$.

Let e_t be the original consumption level (for instance, if the investor bought none of the assets). The following budget constraint states that e_t is split between consumption and investment:

$$c_t = e_t - n_t p_t$$

The second budget constraint states that tomorrow's consumption is increased by the payoff of the investment:

$$c_{t+1}(\omega) = e_{t+1}(\omega) + n_t x_{t+1}(\omega)$$

Theorem 21.1 (Euler's equation).

$$p_t = \beta E_t \left[\frac{u'(c_{t+1})}{u'(c_t)} x_{t+1} \right] \tag{21.1}$$

Proof. The problem is

$$\max_{n_t} u(e_t - n_t p_t) + \beta E_t[u(e_{t+1} + n_t x_{t+1})]$$

Differentiating with respect to n_t we obtain

$$-p_t u'(e_t - p_t n_t) + \beta E_t[u'(e_{t+1} + n_t x_{t+1}) x_{t+1}] = 0$$

or

$$-p_t u'(c_t) + \beta E_t[u'(c_{t+1}) x_{t+1}] = 0 \tag{21.2}$$

■

Equation (21.2) is called *Euler's equation.*

Definition 21.2. *The stochastic discount factor is*

$$m_{t+1}(\omega) = \beta \frac{u'(c_{t+1}(\omega))}{u'(c_t)} \tag{21.3}$$

It is called *stochastic* because it is not known at time t. The stochastic discount factor is also called *marginal rate of substitution, pricing kernel,* or *state price density.* We get rid of utility functions and re-express (21.1) as follows.

Fact 21.3 (pricing equation).

$$p_t = E_t[m_{t+1} x_{t+1}] \tag{21.4}$$

The key observation, as mentioned above, is that the same stochastic discount factor can be used to price every security in the market, as we shall see in the next chapter.

21.2 ▪ Applying the Basic Model

The pricing equation (21.4) $p_t = E_t[m_{t+1} x_{t+1}]$ does *not* depend on the following assumptions (see [21]):

1. markets are complete, or there is a representative investor
2. asset returns are normally distributed

3. investors are two-period investors
4. quadratic utility
5. investors have no capital or labor income.

The first observation is quite important, as financial economists have studied representative investor models for a long time. The idea was to find an investor who aggregates in his beliefs and utility function the behavior of J different investors with potentially different beliefs and preferences. In other words, (21.3) holds for the utility function, subjective discount factor, and subjective probability of the representative investor, as well as aggregate consumption.

To see why (21.4) holds without assuming a representative investor, let us consider a market with only two[65] investors, labeled (I) and (II). Observe that different beliefs are modeled by different expectations, denoted by $E_t^{(I)}$ and $E_t^{(II)}$, or more simply by different probabilities. We define $\pi(\omega_s)$ to be the *objective* (conditional) probability of observing outcome ω_s, while $\pi^{(I)}(\omega_s)$ is the subjective conditional probability that investor (I) attributes to the same outcome, and similarly for $\pi^{(II)}(\omega_s)$. For any random variable x_{t+1}, we have the obvious equality:

$$E_t^{(I)}[x_{t+1}] = \sum_{s=1}^{S} \frac{\pi^{(I)}(\omega_s)}{\pi(\omega_s)} x_{t+1}(\omega_s)\pi(\omega_s) = E_t\left[\frac{\pi^{(I)}}{\pi} x_{t+1}\right] \tag{21.5}$$

The same equation also holds for investor (II) by replacing the superscripts. The first investor solves

$$\max_{n_t^{(I)}} u^{(I)}(c_t^{(I)}) + \beta^{(I)} E_t^{(I)}[u^{(I)}(c_t^{(I)})]$$

while the second investor solves

$$\max_{n_t^{(II)}} u^{(II)}(c_t^{(II)}) + \beta^{(II)} E_t^{(II)}[u^{(II)}(c_t^{(II)})]$$

The first-order conditions for both problems are

$$p_t = \beta^{(I)} E_t^{(I)}\left[\frac{(u^{(I)})'(c_{t+1}^{(I)})}{(u^{(I)})'(c_t^{(I)})} x_{t+1}\right] \tag{21.6}$$

$$p_t = \beta^{(II)} E_t^{(II)}\left[\frac{(u^{(II)})'(c_{t+1}^{(II)})}{(u^{(II)})'(c_t^{(II)})} x_{t+1}\right] \tag{21.7}$$

With the help of (21.5), equations (21.6) and (21.7) are consistent with (21.4), provided (with ω omitted, for clarity)

$$\beta^{(II)} \frac{\pi^{(I)}}{\pi} \frac{(u^{(I)})'(c_{t+1}^{(I)})}{(u^{(I)})'(c_t^{(I)})} = m_{t+1} = \beta^{(II)} \frac{\pi^{(II)}}{\pi} \frac{(u^{(II)})'(c_{t+1}^{(II)})}{(u^{(II)})'(c_t^{(II)})} \tag{21.8}$$

The reader may wonder: what happens if, in practice, equation (21.8) does not hold? It means that equilibrium does not hold. In this case, prices (and consumption streams) should adjust until equilibrium holds. This is the *invisible hand* of the competitive economy, dear to Adam Smith and his followers.

[65] The same development generalizes, of course, to a market with more investors.

21.2.1 ▪ Representation of Equilibrium by Edgeworth Boxes

Readers familiar with (or uninterested in) economics should skip this section. For the others, here is a pictorial representation of equilibrium. Assume we have only two states, i.e., $S = 2$. For simplicity, but without loss of generality, assume that all probabilities (subjective, objective) are equal to $\frac{1}{2}$ and that $(u^{(I)})'(c_t^{(I)}) = (u^{(II)})'(c_t^{(II)}) = 1$. There is an obvious budget constraint, namely, that consumption should be split between the two investors. Since the endowment and the payoff are exogenously given, it is easy to see that there is an exogenously given random variable C_{t+1} that represents aggregate consumption, namely,

$$C_{t+1}(\omega_1) = c_{t+1}^{(I)}(\omega_1) + c_{t+1}^{(II)}(\omega_1) \tag{21.9}$$

$$C_{t+1}(\omega_2) = c_{t+1}^{(I)}(\omega_2) + c_{t+1}^{(II)}(\omega_2) \tag{21.10}$$

Thus the random consumption problem with two states is analogous to a nonrandom allocation problem of two goods, with a fixed quantity of each. Let us label these two goods as widgets and money. The problem is to determine the price of each good. The consumption problem can be described by Figures 21.1(a) to 21.1(d). We represent the isocontours of trader 1 as a function of his quantity of money and widgets and money in Figure 21.1(a). We represent the isocontours of trader 2 as a function of his quantity of money and widgets and money in Figure 21.1(b).

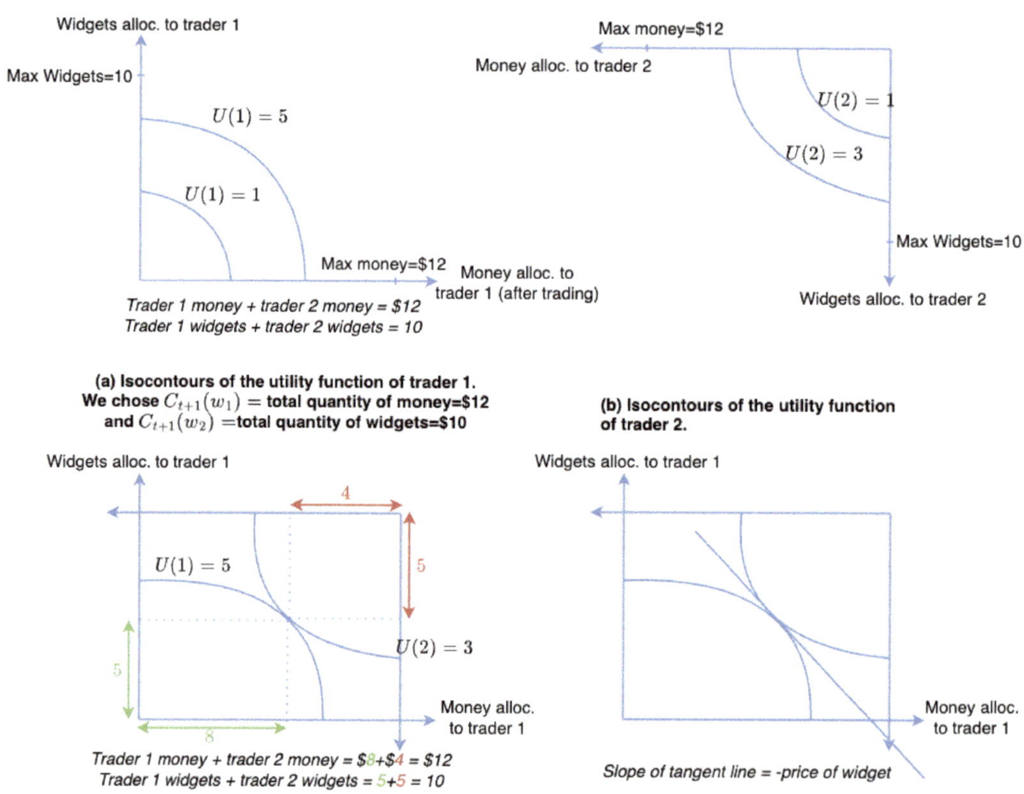

(c) Edgeworth box: Determination of equilibrium allocation (see Exercise 21.1)

(d) Edgeworth box: Determination of equilibrium price (see Exercise 21.1)

Figure 21.1: Representation of equilibrium by Edgeworth boxes.

Since they sum up to (21.9) and (21.10) they can be superposed on Figures 21.1(c) and 21.1(d). Figure 21.1(c) represents the equilibrium allocation, i.e., how traders share widgets and money. Figure 21.1(d) represents the equilibrium price.

Exercises

21.1. (Hard) Figures 21.1(c) and 21.1(d) represent only one point where the gradient of the utility of trader 1 is colinear to the gradient of the utility of trader 2. What information is missing on the Edgeworth box that enables us to pinpoint the equilibrium?

21.2. (Hard) Suppose the risky asset is a stock, which pays at each time t a random dividend $d_t(\omega)$. The payoff of the stock is now

$$x_t(\omega) = p_t(\omega) + d_t(\omega)$$

Now the decision-maker chooses a holding of stocks $n_0, n_1, n_2, \dots$ such that the following budget equation holds:

$$c_{t+1} = e_{t+1} + \underbrace{n_t x_{t+1}}_{\text{revenue of stocks bought at } t} - \underbrace{n_{t+1} p_{t+1}}_{\text{cost of buying stocks at } t+1}$$

The investor's problem at time t is now to maximize

$$E_t\left[\sum_{j=0}^{\infty} \beta^j u(c_{t+j})\right]$$

Show that the price of the risky asset is

$$p_t = E_t\left[\sum_{j=1}^{\infty} \beta^j \frac{u'(c_{t+j})}{u'(c_t)} d_{t+j}\right]$$

21.3. (Hard) Consider the power utility function

$$u(c) = c^{1-\delta}/(1-\delta)$$

where $0 < \delta < 1$ represents the degree of risk aversion.

(a) Show that u is increasing, concave.

(b) Suppose that c_{t+1} is lognormally distributed, i.e.,

$$c_{t+1} = c_t \exp(\sigma z)$$

where $z(\omega)$ is a standard normal random variable. Define $\Delta \ln c_{t+1} = \ln c_{t+1} - \ln c_t$; thus $\Delta \ln c_{t+1}(\omega) = \sigma z(\omega)$. We will show in Chapter 23 that the stochastic discount factor is perfectly negatively correlated with R^T, the return on the tangent portfolio, i.e.,

$$Cov[m_{t+1}, R^T_{t+1}] = -\sqrt{Var_t[R^T_{t+1}]}\sqrt{Var_t[m_{t+1}]}$$

Show that the Sharpe ratio of the tangent portfolio is

$$\frac{E_t[R^T_{t+1}] - R_f}{\sqrt{Var_t[R^T_{t+1}]}} = \sqrt{\exp(\delta^2 Var_t(\Delta \ln c_{t+1})) - 1}$$

(c) Show that it is approximately equal to

$$\delta\sqrt{Var_t(\Delta \ln c_{t+1})}$$

(d) Read the literature on the equity premium puzzle (a Wikipedia article suffices), and conclude that observed Sharpe ratios can be explained only by an abnormally high degree of risk aversion.

Chapter 22

Risk-Neutral Pricing

We now move to a multi-asset world. We observe that the pricing equation (21.4) can be applied to different assets $i = 1 \ldots I$, i.e.,

$$p_{i,t} = E_t[m_{t+1} x_{i,t+1}] \tag{22.1}$$

In other words, the price $p_{i,t}$ at time t of security i is the conditional expected value of the product of the SDF m_{t+1} times the payoff $x_{i,t+1}$. **A single SDF prices all securities.**

22.1 ▪ Consequences of the Pricing Equation

Applying (22.1) to stocks,

$$p_{i,t} = E_t[m_{t+1}(d_{i,t+1} + p_{i,t+1})] \tag{22.2}$$

Dividing this equation by $p_{i,t}$ we obtain for the gross return

$$1 = E_t[m_{t+1} R_{i,t+1}]$$

Applying (22.1) to bonds,

$$p_{0,t} = E_t[m_{t+1}]$$

Dividing this equation by $p_{0,t}$ we obtain for its gross return

$$1 = E_t[m_{t+1} R_f]$$

Rewriting $R_f = 1 + r_f$,

$$1 = E_t[m_{t+1}(1 + r_f)]$$

from which we find that

$$E_t[m_{t+1}] = \frac{1}{1 + r_f}$$

We recover the well-known formula for the price of a one-period bond:

$$p_{0,t} = \frac{1}{1 + r_f}$$

Now, let us define the following random variable:

$$z_{t+1}(\omega) = \frac{m_{t+1}(\omega)}{E_t[m_{t+1}]}$$

Suppose that m_{t+1} is a positive random variable. Then z_{t+1} is also positive with expected value equal to one. We can thus define another probability measure, $\tilde{\mathbb{P}}$, called the **risk-neutral measure**, by

$$\tilde{\mathbb{P}}(\omega) = z_{t+1}(\omega)\mathbb{P}(\omega)$$

Define $\tilde{E}$ to be expectation in the risk-neutral measure, i.e., the measure in which the probability of an outcome ω is given by $\tilde{\mathbb{P}}(\omega)$ and not $\mathbb{P}(\omega)$. It can easily be seen that, for any random variable $y_{t+1}(\omega)$,

$$E_t[z_{t+1}y_{t+1}] = \tilde{E}_t[y_{t+1}]$$

Since $E_t[m_{t+1}] = \frac{1}{1+r_f}$ we can rewrite (22.2) by

$$\begin{aligned} p_{i,t} &= \frac{1}{1+r_f}E_t\left[\frac{m_{t+1}}{E_t[m_{t+1}]}(d_{i,t+1}+p_{i,t+1})\right] \\ &= \frac{1}{1+r_f}E_t[z_{t+1}(d_{i,t+1}+p_{i,t+1})] \end{aligned}$$

We obtain thus the *risk-neutral pricing formula:*

$$p_{i,t} = \frac{1}{1+r_f}\tilde{E}_t[d_{i,t+1}+p_{i,t+1}] \tag{22.3}$$

In other words, the price of an instrument is given by the expected value (in the risk-neutral measure) of its discounted payoffs.

Observe that we can iterate this equation and find

$$p_{i,t} = \tilde{E}_t\left[\sum_{k=1}^{\infty}\frac{1}{(1+r_f)^k}d_{i,t+k}\right] \tag{22.4}$$

22.2 ▪ Option Pricing

A European option can be viewed as an instrument becoming worthless at expiration $T > t$ and giving a payoff only at maturity T. The payoff v_T of the option is a function of the price of another instrument, which we call p_T. For instance, the payoff of a call option is

$$v_T(\omega) = \max(p_T(\omega) - K, 0)$$

where K is a constant called the strike price. Then we can reuse the formula (22.4). Call v_t the price of the call option at time t. Then

$$\begin{aligned} v_t &= \tilde{E}_t\left[\frac{v_T}{(1+r_f)^{(T-t)}}\right] \qquad (22.5) \\ &= \tilde{E}_t\left[\frac{\max(p_T-K,0)}{(1+r_f)^{(T-t)}}\right] \end{aligned}$$

If we adopt a parametric model for the stock price p, we can find an exact formula for the price of the call option v. When stock returns are normal, we obtain the **Black–Scholes** formula for the option price. In the exercise section, you will see the binomial model, where the (scaled) stock return follows the binomial distribution shown in Figure 22.1. You will see in Exercise 22.2 that the expected return of every instrument should be the risk-free rate in the risk-neutral measure. Thus the probability $\tilde{q}$ of the down movement of the stock price has to reflect this constraint. In Exercise 22.3, you will calculate the price of a call option using a binomial model.

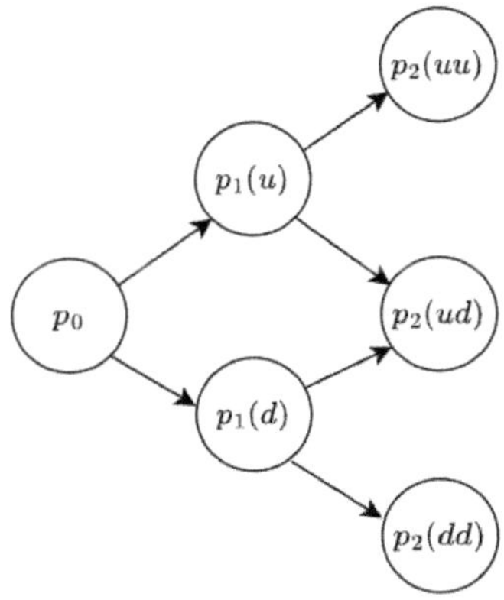

Figure 22.1: Binomial model: the stock price p_1 at time 1 can take two values, the up value $p_1(u)$ and the down value $p_1(d)$.

Exercises

22.1. (Medium) In the multiperiod consumption model from Exercise 21.2, assume that $E_t[\frac{u'(c_{t+j})}{u'(c_t)} d_{t+j}]$ is a constant equal to $\tilde{E}[d]$. Show that the price-earnings ratio of any stock is

$$\frac{p_t}{\tilde{E}[d]} = \frac{\beta}{1-\beta}$$

In this model, dividends do not grow. When they do, we obtain a slightly different formula, which is sometimes called the **Gordon model**.

22.2. (Medium) Show that the expected return of every security "in the risk-neutral measure" is the risk-free rate, that is,

$$\tilde{E}_t[r_{i,t+1}] = r_f \tag{22.6}$$

22.3. In this problem, you will price a call option numerically using formula (22.5). Choose $t = r_f = 0$, $p_0 = 100, K = 105, T = 2$. (Scaled) stock returns follow a binomial process, that is, at each time t,

$$p_{t+1}(\omega) = \begin{cases} 2p_t(\omega) & \text{with probability } 1-\tilde{q} \\ \frac{1}{2}p_t(\omega) & \text{with probability } \tilde{q} \end{cases}$$

(a) Draw a "binomial tree" representing at each node the value taken by $p_{t+1}(\omega)$.

(b) Determine the probability distribution of p_T in the risk-neutral measure by using (22.6), and calculate the expectation (22.5).

(c) Determine recursively the price of the option by first calculating $v_T(\omega) = \max(p_T(\omega) - K, 0)$, then $v_{T-1}(\omega) = \frac{1}{1+r_f}\tilde{E}_{T-1}[\max(p_T(\omega) - K, 0)]$, and so on until time zero. This is called backward induction. Code the program in either MATLAB or Python. Junior programmers should use **For** statements to better understand what is going on. The processing time is immaterial compared to the frustration (and time lost) in understanding vectorization in a programming language.

Chapter 23

The Stochastic Discount Factor

The most important relationship of this book is probably the pricing equation (22.1)

$$p_{i,t} = E_t[m_{t+1} x_{i,t+1}] \tag{23.1}$$

where $p_{i,t}$ is the price at time t of a security i, with payoff at time $t+1$ equal to $x_{i,t+1}$, and m_{t+1} is the stochastic discount factor (SDF). In Chapter 22, we obtained this result with a consumption-based model and proved that the SDF is equal to the ratio of marginal utility of consumptions, multiplied by the subjective discount factor β:

$$m_{t+1}(\omega) = \beta \frac{u'(c_{t+1}(\omega))}{u'(c_t)} \tag{23.2}$$

Relation (23.2) is important for economists who try to understand consumption but is less important for asset managers. For asset managers, the point is that the same SDF prices every security, that is, relation (23.1) holds for every security $i = 0, \ldots, I$, and that the SDF is positive. Thus some securities that are believed to be rightly priced can be used to **calibrate** the model. Once the model is calibrated, it can be used to price other securities.

In this chapter, we show that the pricing equation (23.1) can be derived by the simple law of one price argument. A stronger concept, no-arbitrage, shows that the SDF should be positive. We also provide three different characterizations of the SDF, which have different merit statistically.

23.1 ▪ Payoff Matrix

We first consider only a static problem, so we omit the subscript t. We suppose that we have I securities and S states of the world. The values of the states of the world are labeled $\omega_1, \ldots, \omega_S$. A market is **complete** if an investor can make a bet on the occurrence of every state of the world.

To this effect, we define the payoff matrix

$$\mathbf{A} = \begin{bmatrix} \text{payoff of sec. 1 in state } \omega_1 & \ldots & \text{payoff of sec. 1 in state } \omega_S \\ \vdots & & \vdots \\ \text{payoff of sec. } I \text{ in state } \omega_1 & \ldots & \text{payoff of sec. } I \text{ in state } \omega_S \end{bmatrix}$$

Definition 23.1. *A market is complete*[66] *if the payoff matrix* $\mathbf{A}$ *is square and invertible.*

[66] There are more refined definitions than this one, but we will not consider degenerate cases in this book.

It will be easier in the proof below to give names to the rows of $\mathbf{A}$. We will denote by $\mathbf{a}^1 \in \mathbb{R}^S$ the (row) vector of payoffs of security 1, $\mathbf{a}^2$ the vector of payoffs of security 2, etc.... The payoff a_s^i of security i in state ω_s is given by

$$a_s^i = x_i(\omega_s)$$

Thus the **matrix A** can be rewritten

$$\mathbf{A} = \begin{bmatrix} \mathbf{a}^1 \\ \vdots \\ \mathbf{a}^I \end{bmatrix} = \begin{bmatrix} a_1^1 & \dots & a_S^1 \\ \vdots & & \vdots \\ a_1^I & \dots & a_S^I \end{bmatrix}$$

But the **random payoff vector x** is

$$\mathbf{x}(\omega) = \begin{bmatrix} x_1(\omega) \\ \vdots \\ x_I(\omega) \end{bmatrix}$$

Again, $x_i(\omega)$ represents the payoff of security i in state ω.[67]

Example 23.2. What are **states of the world**? Let us continue with our example of the Claremont Herald from the introduction. Suppose the following four variables are sufficient to explain the price of a newspaper:

- market return
- sectorial (the newspaper industry) return
- amount of interesting local news
- local weather.

Suppose that each explanatory variable can take two values; then there are $2^4 = 16$ states of the world. The first two factors in this example are macroeconomic factors, whereas the last two are firm characteristic factors. The latter impacts readership and thus payoff.[68] Typically, macroeconomic factors are priced, while firm characteristic factors are less important according to modern finance theory.[69]

Let us simplify our example and consider only two explanatory variables: sectorial return and local weather. We will denote by ω a particular state of the world. The following table shows the numbering of the states of the world as well as a (hypothetical) model for the payoff:[70]

State of the world ω	Sectorial return	Weather	Payoff $x(\omega)$
ω_1	High	fair	0.95
ω_2	Low	fair	0.95
ω_3	High	bad	1.1
ω_4	Low	bad	1.1

[67] The reason we have this double notation $\mathbf{a}$ and $\mathbf{x}$ is that it is cleaner to define random variables ($\mathbf{x}$) to calculate expectations.

[68] Indeed, if the news is interesting and local weather is bad, more people are going to read a newspaper.

[69] But not according to the fundamental analyst.

[70] In our model, the payoff x depends on the weather but does not depend on the market return.

We assume constant returns to scale. That is, investing \$1000 in the *Claremont Herald* will return a payoff that is 1000 times what you would get by investing \$1. We can thus normalize and denote by x the payoff by \$1 invested. The pricing formula becomes

$$\text{Value of \$1 invested} = E[mx] \tag{23.3}$$

Denoting by $\pi(\omega_s)$ the probability of state ω_s. (23.3) then becomes[71]

$$\text{Value of \$1 invested in security j} = \sum_{s=1}^{4} \pi(\omega_s)m(\omega_s)x_j(\omega_s) \tag{23.4}$$

What remains to be calculated is the value of the stochastic discount factor $m(\omega_s)$ in all four states of the world ω_s. For this, we take market data. Suppose that we know prices p_j of all four comparable newspapers. Thus we have a system of four equations with four unknowns, which can be solved for the vector $\{m(\omega_s)\}$:

$$\begin{bmatrix} p_1 \\ p_2 \\ p_3 \\ p_4 \end{bmatrix} = \begin{bmatrix} \pi(\omega_1)x_1(\omega_1) & \pi(\omega_2)x_1(\omega_2) & \pi(\omega_3)x_1(\omega_3) & \pi(\omega_4)x_1(\omega_4) \\ \pi(\omega_1)x_2(\omega_1) & \pi(\omega_2)x_2(\omega_2) & \pi(\omega_3)x_2(\omega_3) & \pi(\omega_4)x_2(\omega_4) \\ \pi(\omega_1)x_3(\omega_1) & \pi(\omega_2)x_3(\omega_2) & \pi(\omega_3)x_3(\omega_3) & \pi(\omega_4)x_3(\omega_4) \\ \pi(\omega_1)x_4(\omega_1) & \pi(\omega_2)x_4(\omega_2) & \pi(\omega_3)x_4(\omega_3) & \pi(\omega_4)x_4(\omega_4) \end{bmatrix} \begin{bmatrix} m(\omega_1) \\ m(\omega_2) \\ m(\omega_3) \\ m(\omega_4) \end{bmatrix} \tag{23.5}$$

If the payoffs x_j are normalized, i.e., they are payoff per \$1 invested, then $p_j = 1$ for all j. We then obtain

$$\begin{bmatrix} m(\omega_1) \\ m(\omega_2) \\ m(\omega_3) \\ m(\omega_4) \end{bmatrix} = \begin{bmatrix} \pi(\omega_1)x_1(\omega_1) & \pi(\omega_2)x_1(\omega_2) & \pi(\omega_3)x_1(\omega_3) & \pi(\omega_4)x_1(\omega_4) \\ \pi(\omega_1)x_2(\omega_1) & \pi(\omega_2)x_2(\omega_2) & \pi(\omega_3)x_2(\omega_3) & \pi(\omega_4)x_2(\omega_4) \\ \pi(\omega_1)x_3(\omega_1) & \pi(\omega_2)x_3(\omega_2) & \pi(\omega_3)x_3(\omega_3) & \pi(\omega_4)x_3(\omega_4) \\ \pi(\omega_1)x_4(\omega_1) & \pi(\omega_2)x_4(\omega_2) & \pi(\omega_3)x_4(\omega_3) & \pi(\omega_4)x_4(\omega_4) \end{bmatrix}^{-1} \begin{bmatrix} 1 \\ 1 \\ 1 \\ 1 \end{bmatrix} \tag{23.6}$$

Finally, observe that the stochastic discount factor methodology takes into account both market data and firm-specific information and thus can potentially encompass both fundamental and technical analysis. This conludes Example 23.2.

Voilà. This is the classical theory of asset pricing. In theory, the price of any asset can be solved mechanistically by solving for the stochastic discount factor. In practice, this is much more difficult. The first difficulty is that typically markets are **incomplete**, meaning that the number of states of the world S is larger than the number I of liquidly traded securities: too few equations, too many unknowns. The pricing problem is then **underdetermined**. Our goal is then to augment the number of pricing equations.[72]

A solution is to make use of data (such as newspaper prices) at different times $t \in \{1, \ldots, T\}$, as we shall see in Chapter 23. The number of equations is then $I * T$. This can be made larger than the number of states of the world if one reduces the functional form of our model for the SDF. For instance, one can assume stationarity, i.e., that the collection of random variables $\{m_t\}$ is IID.

[71] In this example, sectorial return does not depend on geography but does depend on the weather. As a result, the probabilities are city-specific. This detail is not too important. The important thing to remember is that probabilities can be determined easily.

[72] The pricing problem may become overdetermined, but it is not an issue. Rather than requiring exact pricing of every liquid security, an academic will take as an objective to minimize the mispricing. The practically minded individual will just buy low and sell high.

23.2 ▪ Existence of the SDF by the Law of One Price

We recall the following notation:

- There are I securities labeled $1, \ldots, i, \ldots, I$.
- There are S states labeled $\omega_1, \ldots, \omega_s, \ldots, \omega_S$. The sample space is $\Omega = \{\omega_1, \ldots, \omega_S\}$.

The payoff space $\mathcal{X}$ is the set of all payoffs x, viewed as random variables, that investors can purchase.

We make the following assumptions.

Assumption 2. $x_1, x_2 \in \mathcal{X}$ implies that $ax_1 + bx_2 \in \mathcal{X}$ for any real a, b.

This assumption says that if an investor can buy one share of the stock, she can also buy 10 shares of the stock. She can also sell the stock short. Any portfolio can be made.

Let us now call $p : \mathcal{X} \rightarrow \mathbb{R}$ the pricing functional. It assigns to each payoff x a price p.

Assumption 3 (law of one price, linearity). For $a, b \in \mathbb{R}$, $p(ax_1 + bx_2) = ap(x_1) + bp(x_2)$.

This assumption says that the price of a lunch made of an entree and a drink is the sum of the price of the entree and the price of the drink. The price of two drinks is twice the price of a drink. There are no discounts on quantity.

Table 23.1 below should clarify what variable is a vector and what variable is a random vector.

Vector	Corresponding random variable
Generic security payoff $\mathbf{a} \in \mathbb{R}^I$	$x \in \mathcal{X} : \Omega \to \mathbb{R}$ with $x(\omega_s) = a_s$
Traded security payoff $\mathbf{a}^i \in \mathbb{R}^I$	$x_i \in \mathcal{X} : \Omega \to \mathbb{R}$ with $x_i(\omega_s) = a_s^i$
Completeness: there exists $\mathbf{c} \in \mathbb{R}^I$ that satisfies $\mathbf{a}^* = \sum_i c_i \mathbf{a}^i$	If $x^*(\omega_s) = a_s^*$ the same $\mathbf{c} \in \mathbb{R}^I$ satisfies $x^*(\omega_s) = \sum_i c_i x_i(\omega_s)$

Table 23.1: Notation for vectors and random variables.

We concatenate these into vectors:

$$\mathbf{p} = \begin{bmatrix} p(x_1) \\ \vdots \\ p(x_i) \\ \vdots \\ p(x^I) \end{bmatrix} \in R^I \quad \mathbf{x}(\omega) = \begin{bmatrix} x_1(\omega) \\ \vdots \\ x_i(\omega) \\ \vdots \\ x_I(\omega) \end{bmatrix}$$

Theorem 23.3. *Given Assumptions 2 and 3, there exists a payoff $x^* \in \mathcal{X}$ such that*

$$p(x) = E[x^* x] \text{ for all } x \in \mathcal{X} \tag{23.7}$$

If $\mathbf{x}$ spans $\mathcal{X}$, then

$$x^*(\omega) = \mathbf{p}' E[\mathbf{x}\mathbf{x}']^{-1} \mathbf{x}(\omega) \tag{23.8}$$

Observation 1. Again, the key part of this theorem is that the same discount factor x^* prices every asset x. This opens the door to *relative* pricing: if we observe the prices $p(x_1)$ and $p(x_2)$ of assets 1 and 2, and want to price asset 3, for which no price is given, we can reuse the discount factor x^* calculated using x_1 and x_2 to calculate the price of the third asset $p(x_3)$.

Observation 2. You may wonder why we wrote x^* and not m. The reason is that the SDF may not be unique if the markets are not complete. So there may be several formulas for m, but only one for x^*. In the incomplete market case, the SDF is any random variable m with

$$m(\omega) = x^*(\omega) + \varepsilon(\omega)$$

for any random variable ε uncorrelated with $x \in \mathcal{X}$ (i.e., $E[x\varepsilon] = 0$). Thus, x^* is the projection of any SDF $m(\omega)$ on $\mathcal{X}$, which we denote by

$$x^* = proj(m|\mathcal{X})$$

Note that ε corresponds to the *idiosyncratic risk* referred to above.

Observation 3. The same "ingredients" are in formulae (23.8) and (23.6): the payoff $\mathbf{x}(\omega)$ and the security prices $\mathbf{p}$. Observe that

- in (23.6), we need to know the state probabilities $\pi(\omega)$
- in (23.8), we need to know only the second moments.

In practice, estimators of second moments work better than estimators of probabilities, so (23.8) is better.

But now that we know under what hypothesis an SDF exists, there is a much easier, equivalent way to calculate it. We will see it in the next section.

Proof. We give a proof only in the complete market case. The incomplete case is handled in exercises at the end of the chapter. An Arrow–Debreu security (for state ω_s) is a security that pays \$1 if state ω_s occurs and zero otherwise. If the market is complete, all the Arrow–Debreu securities belong to $\mathcal{X}$, so that the payoff x_s of each Arrow–Debreu security is

$$x_s(\omega) = 1[\omega = \omega_s] = \begin{cases} 1 & \text{if } \omega = \omega_s \\ 0 & \text{otherwise} \end{cases}$$

Let us call $z_s \equiv p(x_s)$ the price of the Arrow–Debreu security for state s, or, in short, the **state price**. By definition, x^*, if it exists, must satisfy

$$z_s = \sum_s \pi(\omega_s) x^*(\omega_s) 1[\omega = \omega_s]$$

The latter is clearly equal to

$$z_s = \pi(\omega_s) x^*(\omega_s)$$

By Assumption 2, there exists a payoff $x^* \in \mathcal{X}$ such that

$$x^*(\omega_s) = \frac{z_s}{\pi(\omega_s)}$$

Any payoff x_j is a combination of payoffs of Arrow–Debreu securities:

$$x_j(\omega) = \sum_s x_j(\omega_s) 1[\omega = \omega_s]$$

Thus, by Assumption 3 (linearity of pricing),

$$p(x_j) = \sum_s x_j(\omega_s)p(1[\omega = \omega_s]) \tag{23.9}$$

$$= \sum_s x_j(\omega_s)z_s \tag{23.10}$$

$$= \sum_s x_j(\omega_s)x^*(\omega_s)\pi(\omega_s) \tag{23.11}$$

$$= E[x^* x_j] \tag{23.12}$$

Thus existence is proved. Since $x^* \in \mathcal{X}$ it can be represented as a combination of traded security payoffs:

$$x^*(\omega_s) = \sum_i c_i x_i(\omega_s)$$

Thus we can develop (23.12) as

$$p(x_j) = \sum_s \pi(\omega_s)x^*(\omega_s)x_j(\omega_s)$$

$$= \sum_s \pi(\omega_s)\left(\sum_i c_i x_i(\omega_s)\right) x_j(\omega_s)$$

Permuting summations on s and i, we obtain

$$p(x_j) = \sum_i \left(\sum_s \pi(\omega_s)x_i(\omega_s)x_j(\omega_s)\right) c_i \tag{23.13}$$

$$= \sum_i E[x_i x_j]c_i$$

Doing that for every security, we have

$$\begin{bmatrix} p(x_1) \\ \vdots \\ p(x_j) \\ \vdots \\ p(x_I) \end{bmatrix} = \begin{bmatrix} E[x_1x_1] & \dots & E[x_ix_1] & \dots & E[x_Ix_1] \\ \vdots & & \vdots & & \vdots \\ E[x_1x_j] & \dots & E[x_ix_j] & \dots & E[x_Ix_j] \\ \vdots & & \vdots & & \vdots \\ E[x_1x_I] & \dots & E[x_ix_I] & \dots & E[x_Ix_I] \end{bmatrix} \begin{bmatrix} c_1 \\ \vdots \\ c_i \\ \vdots \\ c_I \end{bmatrix}$$

or, with the obvious notation,

$$\mathbf{p} = E[\mathbf{x}\mathbf{x}']\mathbf{c}$$

from which we obtain

$$\mathbf{c}' = \mathbf{p}'E[\mathbf{x}\mathbf{x}']^{-1}$$

We can now rewrite $\mathbf{a}^* = \sum_i c_i \mathbf{a}^i$ as $\mathbf{a}^* = \mathbf{c}'\mathbf{A}$. Thus

$$\mathbf{a}^* = \mathbf{p}'E[\mathbf{x}\mathbf{x}']^{-1}\mathbf{A}$$

from which we find

$$\begin{aligned} x^*(\omega_s) &= a_s^* \\ &= (\mathbf{p}'E[\mathbf{x}\mathbf{x}']\mathbf{A})_s \\ &= \left(\mathbf{p}'E[\mathbf{x}\mathbf{x}']^{-1} \begin{bmatrix} a_1^1 & \dots & a_s^1 & \dots & a_S^1 \\ \vdots & & \vdots & & \vdots \\ a_1^i & \dots & a_s^i & \dots & a_S^i \\ \vdots & & \vdots & & \vdots \\ a_1^I & \dots & a_s^I & \dots & a_S^I \end{bmatrix} \right)(s) \end{aligned}$$

This means that we take the sth column of the $\mathbf{A}$ matrix, i.e.,

$$x^*(\omega_s) = \mathbf{p}'E[\mathbf{x}\mathbf{x}']^{-1} \begin{bmatrix} a_s^1 \\ \vdots \\ a_s^i \\ \vdots \\ a_s^I \end{bmatrix}$$

Again, since $a_s^i = x_i(\omega_s)$, we have

$$x^*(\omega_s) = \mathbf{p}'E[\mathbf{x}\mathbf{x}']^{-1} \begin{bmatrix} x_1(\omega^s) \\ \vdots \\ x_i(\omega^s) \\ \vdots \\ x_I(\omega^s) \end{bmatrix}$$

■

The astute reader will recognize in this latest formula one (which one?) of the Moore–Penrose inverses of the $\mathbf{A}$ matrix; see section 6.1.1.

23.3 ▪ Positivity of the SDF by No-Arbitrage

Definition 23.4. *Let* $\mathbf{1}$ *be the vector of ones. An arbitrage portfolio is a row vector* $\mathbf{w} \in \mathbb{R}^I$ *such that*

$$\mathbf{w}\mathbf{A} \geq \mathbf{0} \tag{23.14}$$

$$\mathbf{w}\mathbf{1} < \mathbf{0} \tag{23.15}$$

In other words, the portfolio has a negative value at time zero when you buy the securities and a positive payoff at time 1 in all states of the world. This definition can be reformulated as follows and extended to the payoff space.

Definition 23.5. *A payoff space* $\mathcal{X}$ *and pricing function* $p(x)$ *leave no arbitrage opportunities if every payoff* $x(\omega)$ *that is always nonnegative, i.e.,* $P(x \geq 0) = 1$, *and sometimes positive, i.e.,* $P(x > 0) > 0$, *has positive price* $p(x) > 0$.

Obviously, if $m(\omega) > 0$, then $p(x) > 0$ for positive x, the positivity of the discount factor implies the absence of arbitrage. The converse proposition is more interesting.

Theorem 23.6 (no-arbitrage). *In complete markets, no-arbitrage and the law of one price imply that there exists a unique $m > 0$ such that $p = E[mx]$.*

Proof. Suppose that $m(\omega_i) \leq 0$ for some states ω_i with $i \in \mathcal{I} \subset [1, \ldots, S]$. Form a payoff $x(\omega)$ which is equal to one if $\omega = \omega_i$ with $i \in \mathcal{I}$ and zero otherwise. This payoff is strictly positive, but its price

$$p(x) = \sum_{i \in \mathcal{I}} \pi(\omega_i) m(\omega_i)$$

is negative or zero, negating the assumption of no-arbitrage. ■

When markets are incomplete, the proof is harder. You will see in the exercise section two problems that guide you through two slightly different proofs. The first one is geometric. When $S = 1$, the picture given by equations (23.14), (23.15) can be represented in the plane, with an axis equal to payoff at time 0 and another axis equal to payoff at time 1. Obviously, both cannot be positive if there is no arbitrage (see Figure 23.1). The second proof is more algebraic. It is an application of duality in linear programming.

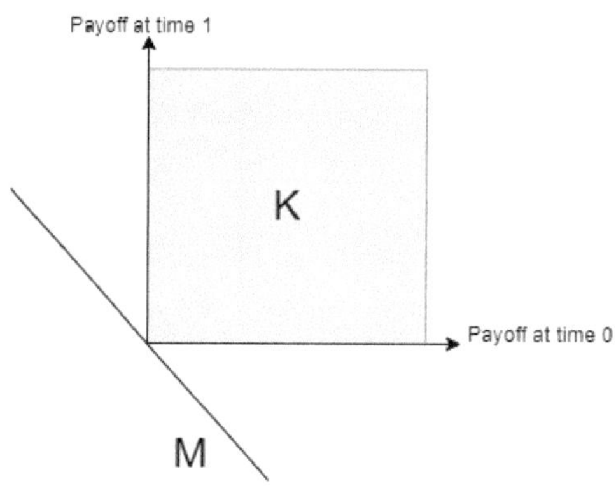

Figure 23.1: The subspace M represents (payoff at time 0, payoff at time 1) under no-arbitrage: if one is positive, the other one is negative. The subspace K is the positive quadrant (positive payoffs at times 0 and 1).

23.4 ▪ Relation between the SDF and Excess Returns Moments

We now return to a multiperiod model in order to better understand the structure of the SDF. For simplicity, we assume that markets are complete, but all relations below also hold if markets are not complete. Thus there is a unique SDF m such that $E_t[m_{t+1}] = 1/R^f$ and $E_t[x_{i,t+1} m_{t+1}] = p_t$ for any payoff $x_{i,t+1}$. It is equal to x^*. In what follows, we will get rid of the normalization relation $E_t[m_{t+1}] = 1/R^f$, which is not necessary when we price excess returns.

Definition 23.7. *The excess return $r^e_{i,t+1}$ of security i is*

$$r^e_{i,t+1} = r_{i,t} - r_f \tag{23.16}$$

The no-arbitrage relation (valid for any i) is

$$E_t[r^e_{t,i} m_{t+1}] = 0 \tag{23.17}$$

Let $M_{t+1} = c_t m_{t+1}$ for any constant $c_t \neq 0$. Then obviously

$$E_t[r^e_{t+1,i} m_{t+1}] = 0 \Longleftrightarrow E_t[r^e_{t+1,i} M_{t+1}] = 0 \tag{23.18}$$

Thus we have one degree of freedom in choosing the SDF. Many authors choose the following one, for practical reasons.

Definition 23.8. *The SDF for excess returns is the random variable M_{t+1} such that*

$$M_{t+1} = 1 - F_{t+1} \tag{23.19}$$

where

$$F_{t+1} = (\mathbf{r}^e_{t+1})^t (E_t[\mathbf{r}^e_{t+1} (\mathbf{r}^e_{t+1})^t])^{-1} E_t[\mathbf{r}^e_{t+1}]$$

Theorem 23.9. *The SDF for excess returns M_{t+1} satisfies the no-arbitrage condition (23.18) for every security i.*

Proof. We calculate

$$\begin{aligned}
E_t[\mathbf{r}^e_{t+1} M_{t+1}] &= E_t[\mathbf{r}^e_{t+1}(1 - F_{t+1})] \\
&= E_t[\mathbf{r}^e_{t+1}] - E_t[\mathbf{r}^e_{t+1}(\mathbf{r}^e_{t+1})^t (E_t[\mathbf{r}^e_{t+1}(\mathbf{r}^e_{t+1})^t])^{-1}] E_t[\mathbf{r}^e_{t+1}] \\
&= E_t[\mathbf{r}^e_{t+1}] - E_t[\mathbf{r}^e_{t+1}] \\
&= \mathbf{0}
\end{aligned}$$

■

23.5 ▪ Relation between the SDF and the Tangency Portfolio

We found relation (16.16) for the tangent portfolio weights $\mathbf{w}^T$. Here we will find a different relation, which will prove to be easier for interpreting the SDF. Recall that r^p_t is the return on any portfolio. The one-period mean-variance approach consists in minimizing the variance of r^p under a constraint on the mean of r^p:

$$\begin{aligned}
&\min_{\mathbf{w}} Var[r^p] \\
&E[r^p] = \mu
\end{aligned} \tag{23.20}$$

Since the variance of a constant is zero, the solution to this problem is the same as the solution to the following problem:

$$\begin{aligned}
&\min_{\mathbf{w}} Var[r^p - r^f] \\
&E[r^p - r_f] = \mu - r_f
\end{aligned} \tag{23.21}$$

We use our excess return notation:

$$\begin{aligned}
&\min_{\mathbf{w}} Var[r^{e,p}] \\
&E[r^{e,p}] = \mu - r_f
\end{aligned} \tag{23.22}$$

Take μ to be the expected return on the tangent portfolio. Since $\mu - r^f$ is constant, we may as well write the problem above as

$$\min_{\mathbf{w}} Var[r^{e,p}] + (E[r^{e,p}])^2 \qquad (23.23)$$
$$E[r^{e,p}] = \mu - r_f$$

One recognizes $Var[r^{e,p}] + (E[r^{e,p}])^2$ as the second moment of the excess portfolio return $r^{e,p}$. Since $r^{e,p} = \mathbf{w}'\mathbf{r}^e$, the latter is

$$E[(r^{e,p})^2] = \mathbf{w}' E[\mathbf{r}^e(\mathbf{r}^e)']\mathbf{w}$$

The Lagrangian approach returns

$$\mathbf{w}^* = \gamma(E[\mathbf{r}^e(\mathbf{r}^e)'])^{-1}E[\mathbf{r}^e]$$

Now let us return to our multiperiod model. Denoting by $\tilde{\omega}_t$ the weight,

$$\tilde{\omega}_t = (E_t[\mathbf{r}^e_{t+1}(\mathbf{r}^e_{t+1})^t])^{-1}E_t[\mathbf{r}^e_{t+1}] \qquad (23.24)$$

By definition, weights in a portfolio must add up to one. Thus the tangent portfolio is

$$\mathbf{w}_t^T = \gamma_t\tilde{\omega}_t$$

where, as in the single-period model, $1/\gamma_t$ can be interpreted as the degree of risk aversion. Thus we have the following important result.

Theorem 23.10. *The random variable F_{t+1} in (23.19) is a scaled version of the tangent portfolio return*

$$F_{t+1} = \frac{(\mathbf{w}^T)'\mathbf{r}^e_{t+1}}{\gamma_t}$$

This theorem reasserts one of the main messages of this book. Asset pricing is very closely linked to asset management, as Figure 23.2 illustrates.

We shall see in Chapter 25 the generalized method of moments approach to estimate the SDF and in Chapter 26 the neural network version of it. It bypasses completely the inversion of the second moment of the return matrix.

Exercises

23.1. (Easy) Let the payoff matrix of two securities be

$$\mathbf{A} = \begin{bmatrix} 4/3 & 2/3 \\ 1/3 & 4/3 \end{bmatrix}$$

In other words, the payoff of security 1 in states 1 and 2 is

$$x_1(\omega_1) = A_{1,1} = 4/3$$
$$x_1(\omega_2) = A_{1,2} = 2/3$$

while the payoff of security 2 is

$$x_2(\omega_1) = A_{2,1} = 1/3$$
$$x_2(\omega_2) = A_{2,2} = 4/3$$

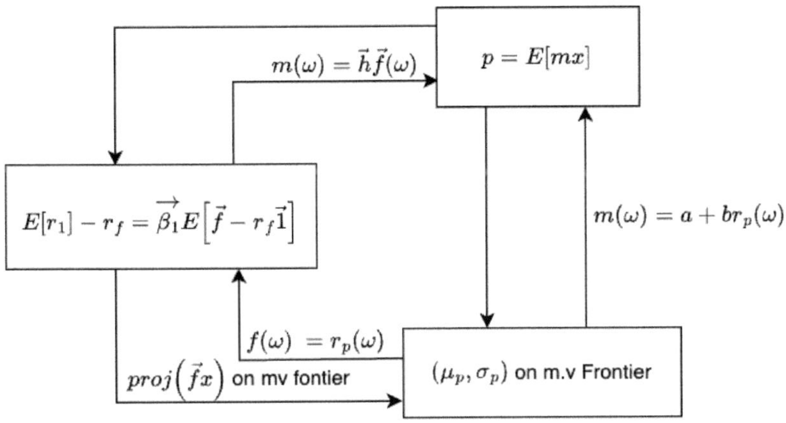

Figure 23.2: Relationship between different concepts in pricing. The law of one price implies the existence of the SDF. The SDF is an affine function of any portfolio return on the mean-variance frontier. It is a linear function of the factors.

Suppose that the state probabilities are

$$\pi=\begin{bmatrix}3/4\\1/4\end{bmatrix} \tag{23.25}$$

Let the price of the securities be

$$\mathbf{p}'=\begin{bmatrix}1 & 1\end{bmatrix}$$

You should calculate the discount factor $m(\omega)$ in each state ω according to

(a) formula (23.6) in section 23.1

(b) formula (23.8) in section 23.2.

23.2. This is the same problem as Exercise 23.1, except that you do not have a parametric model to describe the data. Use Monte Carlo with T =100 scenarios to simulate outcomes according to the distribution (23.25). Suppose that each outcome corresponds to a different time. You now have two time series, one for each security, namely, $\{x_{t,1}\}$ and $\{x_{t,2}\}$ for $t=1,\ldots,100$ of payoffs. For each $i=1,2$, $x_{t,i}$ is IID. Forget the distribution you sampled from. You are now in the situation of an econometrician who needs to analyze a historical time series. Assume that the discount factor does not depend on time. You should calculate the discount factor $m(\omega)$ in each state ω according to

(a) formula (23.6) in section 23.1, using empirical probabilities

(b) formula (23.8) in section 23.2, using estimated second moments

(c) the gradient method, where the function to minimize is

$$\begin{aligned}\min_{m_1,m_2}&\sum_{t=1}^{T}((x_1(\omega_1)m_1-x_1(\omega_2)m_2)1\{x_{t,1}=x_1(\omega_1)\}+x_1(\omega_2)m_2-1)^2\\&+\sum_{t}^{T}((x_2(\omega_1)m_1-x_2(\omega_2)m_2)1\{x_{t,2}=x_2(\omega_1)\}+x_2(\omega_2)m_2-1)^2\end{aligned}$$

(d) Discuss the results. If conclusions are not clear, increase T.

23.3. (Hard) In this problem, you will prove that either there exists an arbitrage opportunity or there exists a risk-neutral measure. The proof is by linear programming and does not make the assumption that the payoff matrix $\mathbf{A}$ is square: there can be more states than securities (a generically incomplete market) or less.

(a) Using the definition in Chapter 5, show that the dual of the problem

$$\max_{\mathbf{m}} \mathbf{0} * \mathbf{m}$$
$$\mathbf{Am} = \mathbf{1}$$
$$\mathbf{m} \geq \mathbf{0}$$

is

$$\min_{\mathbf{w}} \mathbf{w1} \tag{23.26}$$
$$\mathbf{wA} \geq \mathbf{0}$$

(b) By normalizing the state price $\mathbf{m}$ if necessary, define a proper risk-neutral measure, and show that either there exists a risk-neutral measure or there exists an arbitrage portfolio.

23.4. (Hard) As in Exercise 23.3, you will prove that either there exists an arbitrage opportunity or there exists a risk-neutral measure. The proof is geometric. Use the same notation as in Exercise 23.1. Define $M = \{-\mathbf{y1}, \mathbf{yA}\} \in \mathbb{R} \times \mathbb{R}^S$ and $K = \mathbb{R}_+ \times \mathbb{R}^S_+$.

(a) Take $S = 1$. Refer to Figure 23.1 and show that if there is no arbitrage, then $M \cap K = \{0\}$.

(b) Take $S = 2$. Graph M and K.

(c) Take a general S. The separating hyperplane theorem implies the existence of a linear functional $F : \mathbb{R} \times \mathbb{R}^S \to \mathbb{R}$ such that $F(\mathbf{u}) < F(\mathbf{x})$ for all $\mathbf{u}$ in M and nonzero $\mathbf{x}$ in K. Since M is a linear space, this implies that $F(\mathbf{u}) = 0$ for all $\mathbf{u}$ in M and that $F(\mathbf{x}) > 0$ for all nonzero $\mathbf{x}$ in K. Thus, for any $\mathbf{x} \equiv (v, \mathbf{c})$, F can be represented as $F(\mathbf{x}) = \alpha v + \mathbf{c}'\psi$ for some $\alpha > 0$ in $\mathbb{R}$ and some $\psi \gg \mathbf{0}$ in $\mathbb{R}^S$.

(d) Suppose there is no arbitrage. What is $F(-\mathbf{y1}, \mathbf{yA})$ equal to?

(e) Define m as a function of α and ψ, and show that it is positive if there is no arbitrage. Hint: If you cannot do it by yourself, look at [68].

Chapter 24

Estimation of Linear Factor Models of Return

The workhorse of the asset management industry has been (and to some extent still is) linear factor models. In 1964, Sharpe [70] published the CAPM. Fama and French developed a three-factor model [72] and then a five-factor model [73].[73]

The idea of the CAPM was that assets that lose value when the single factor, i.e., the overall stock market, goes down are risky and must give their owners risk premia. Conversely, "safe assets," like gold or the Swiss franc, should earn a lower (or negative) risk premium.

A common view in academia is that assets that will do badly in bad times will earn large positive risk premia. The CAPM specifies these bad times as times of low market returns, but some investors/households have different definitions of bad times, which somehow justify the use of factors other than the overall market return. One of the main proponents of the efficient market theory, Eugene Fama [57], dubbed this view a "fishing license," i.e., anybody can fish for any factor, provided it gives good results. The view of Campbell and Yogo [71] is that "the best hope for finding robust factors ... is to try to understand the fundamental macroeconomic sources of risk." As this is a book on machine learning, we will not go into economic modeling, and we will let the machine identify the factors.

Definition 24.1. *A linear K-factor model of the excess return r_i^e of firm i has the form*

$$r_i^e(t+1,\omega) = \alpha_i + \sum_{k=1}^{K} \beta_{i,k} f_k(t) + u_i(t+1,\omega) \tag{24.1}$$

The stochastic processes $f_k(t)$ are called factors, and u is called the noise. The noise is supposed to be mean-zero.

If we know the identity of the factors, like in the Fama–French model, we can estimate the factor model by regression. Often, the identity of the factor is unknown, and one can resort to different techniques, often referred to as *factor analysis*. Principal component analysis is one of them (see Chapter 10).

Definition 24.2. *Suppose factors are ergodic processes. The term $\lambda_k \equiv E[f_k(t)]$ is called the market price of risk of factor k. The term $\beta_{i,k}\lambda_k$ is called the risk premium (of factor k).*

[73]Although one, the small stock premium, has been fading with time. This is a perfect illustration that markets often learn, and what worked 30 years ago may not work now. Some strategies seem to endure, like short volatility strategies, momentum, and value investing.

At first glance, equation (24.1) is just a linear regression of a time series, r_i^e, on other time series, the factors. Few economists would call it a pricing *model*. There is, however, an important point. The law of one price stipulates that **if the factors are excess returns**[74] **and markets are efficient, alpha should be zero**.

The condition that factors should be excess returns is perhaps not so important in practice. It is not that hard to find portfolios that are highly correlated with a specific factor.

As mentioned earlier, there is a subtle drawback in the model (24.1), even with the proviso that alphas should be zero in an efficient market. It does not imply the absence of arbitrage. As markets become more and more arbitrage-free, factor models of returns should lose their attractiveness. It is our (untested) opinion that ultimately SDF models, which will be covered in the next chapter, should prevail over the models from this chapter.

For machine learning practitioners, the linearity of the model is not particularly important. We shall see **nonlinear models** of the SDF in the next chapter and go back to nonlinear models of returns in Chapters 26 and 27.

How can we estimate the market price of risk? If we assume ergodicity, then a good estimator would be the sample mean:

$$\lambda_k = E_T[f_k]$$

where we used Hansen's time-average symbol E_T, introduced in Chapter 8. However, the market price of risk can be time-varying. A commonly held view is that betas are more stable than the market price of risk, in particular in times of crisis. Thus, better estimation techniques have been developed. We will mention

- the BARRA method (with GLS correction)
- GMM
- Kalman filter approach.[75]

All these methods are moment-based approaches. For robust approaches, see Bailer, Maravina, and Martin [74].

Before turning to these approaches, we look at successive attempts to solve this problem through a simple numerical example.

24.1 ▪ Example

We have $I = 2$ securities and $T = 2$ times. Organize securities by rows:

$$\{r_{i,t}^e\} = \begin{bmatrix} 1.2 & 1.6 \\ 0.8 & 1.0 \end{bmatrix} \tag{24.2}$$
$$f = \begin{bmatrix} 0.1 & 0.2 \end{bmatrix}$$

For ease of exposition, we assume that $\alpha = 0$. We start with a model with time-varying betas:

$$r_{i,t+1}^e = \beta_{i,t} f_t + u_{i,t+1} \tag{24.3}$$

Based on this data, we will attempt to estimate the betas and the risk premium. Our **first attempt** is to assume that expected returns are martingales. In that case, taking conditional

[74]If not, all intercepts α_i should take the same value.

[75]Which can also be used to estimate betas.

expectations, namely, $E_t[r^e_{i,t+1}] = r^e_{i,t}$, yields the two relations

$$\begin{bmatrix} 1.2 \\ 0.8 \end{bmatrix} = \begin{bmatrix} \beta_{1,1} \\ \beta_{2,1} \end{bmatrix} \lambda \quad \text{for } t = 1$$
$$\begin{bmatrix} 1.6 \\ 1.0 \end{bmatrix} = \begin{bmatrix} \beta_{1,2} \\ \beta_{2,2} \end{bmatrix} \lambda \quad \text{for } t = 2$$

which are underidentified, unless we take $\lambda = E_T[f] = 0.15$, the time-average of the factors. But this trivial estimator does not satisfy our goal, which is to find the best possible estimator of the risk premium. In our **second attempt**, we take constant betas, i.e.,

$$r^e_{i,t+1} = \beta_i f_t + u_{i,t+1}$$

We then aggregate returns over time and obtain in our example

$$\begin{bmatrix} E_T[r^e_1] \\ E_T[r^e_2] \end{bmatrix} = \begin{bmatrix} 1.4 \\ 0.9 \end{bmatrix} = \begin{bmatrix} \beta_1 \\ \beta_2 \end{bmatrix} \lambda \tag{24.4}$$

This is still unidentified (unless we make an assumption on λ like before). In our **third** (and first successful) **attempt** we first do a time series regression to find the beta of each stock:

$$\begin{bmatrix} 1.2 & 1.6 \end{bmatrix} = \beta_1 \begin{bmatrix} 0.1 & 0.2 \end{bmatrix} + \text{noise} \Rightarrow \beta_1 = 8.8$$
$$\begin{bmatrix} 0.8 & 1.0 \end{bmatrix} = \beta_2 \begin{bmatrix} 0.1 & 0.2 \end{bmatrix} + \text{noise} \Rightarrow \beta_2 = 5.6$$

and then do a cross-sectional regression:

$$\begin{bmatrix} E_T[R^e_1] \\ E_T[R^e_2] \end{bmatrix} = \begin{bmatrix} \beta_1 \\ \beta_2 \end{bmatrix} \lambda + \text{noise} \tag{24.5}$$

This is exactly like (24.4), except that now the β_i are identified:

$$\begin{bmatrix} 1.4 \\ 0.9 \end{bmatrix} = \begin{bmatrix} 8.8 \\ 5.6 \end{bmatrix} \lambda + \text{noise}$$

from which we calculate

$$\lambda = 0.16$$

This is the method adopted by the software manufacturer BARRA. We can extend it to find a time-dependent risk premium. This **fourth attempt** is the Fama–MacBeth [75] approach.

$$\begin{bmatrix} r^e_{1,t} \\ r^e_{2,t} \end{bmatrix} = \begin{bmatrix} \beta_1 \\ \beta_2 \end{bmatrix} \lambda_t + \text{noise} \tag{24.6}$$

In our example,

$$\begin{bmatrix} 1.2 \\ 0.8 \end{bmatrix} = \begin{bmatrix} 8.8 \\ 5.6 \end{bmatrix} \lambda_1 + \text{noise}$$
$$\begin{bmatrix} 1.6 \\ 1.0 \end{bmatrix} = \begin{bmatrix} 8.8 \\ 5.6 \end{bmatrix} \lambda_2 + \text{noise}$$

24.2 ▪ Determination of the Price of Risk by the BARRA Method

The BARRA method is a two-step method. First, do a cross-sectional regression, then a time series regression:

$$\mathbf{r}^e_{t+1} = \boldsymbol{\alpha} + \boldsymbol{\beta} f_t + \mathbf{u}_{t+1} \tag{24.7}$$

The market price of risk λ, if supposed ergodic, then satisfies

$$E_T(\mathbf{r}^e) - \hat{\boldsymbol{\alpha}} \simeq \hat{\boldsymbol{\beta}}\lambda \tag{24.8}$$

The following algorithm uses a regression instead.

ALGORITHM 24.1. Risk-premium BARRA.

```
1.  Given time series {r_t^e}, {f_t},
2.  Determine estimates (α̂,β̂) of (α,β) by regression of (24.7)
3.  Determine λ by regressing r^e − α̂ on β̂
```

The solution by OLS is obviously

$$\hat{\lambda} = (\hat{\boldsymbol{\beta}}'\hat{\boldsymbol{\beta}})^{-1}\hat{\boldsymbol{\beta}}' E_T(\mathbf{r}^e)$$

If we use GLS, we obtain

$$\hat{\lambda} = (\boldsymbol{\beta}'\boldsymbol{\Sigma}^{-1}\boldsymbol{\beta})^{-1}\boldsymbol{\beta}'\boldsymbol{\Sigma}^{-1} E_T(\mathbf{r}^e) \tag{24.9}$$

where $\boldsymbol{\Sigma}$ is the variance-covariance matrix of $\mathbf{u}$. It turns out that GLS is no longer the optimal estimate.

24.3 ▪ Determination of the Price of Risk by GMM

Rather than a two-step method, we can formulate the problem in one step by GMM. The parameter vector is now

$$\boldsymbol{\theta}' = [\ \mathrm{a}' \quad \mathrm{b}' \quad l\ ]$$

where $\mathbf{a}$ stands for alpha, $\mathbf{b}$ for beta, and l for lambda. A possible (non optimal) choice for the moment condition is:

$$\mathrm{g}_T(\boldsymbol{\theta}) = \begin{bmatrix} E_T[\mathbf{r}^e - \mathbf{a} - \mathbf{b}f] \\ E_T[(\mathbf{r}^e - \mathbf{a} - \mathbf{b}f)f] \\ E_T[(\mathbf{r}^e - \mathbf{a} - \mathbf{b}l)'\mathbf{b}] \end{bmatrix}$$

24.4 ▪ Kalman Filter Estimation

The Kalman filter is a recursive least squares algorithm to estimate an unobservable (hidden) stochastic process, called the *state process* when the noise is Gaussian. We will not give a full exposition of the assumptions behind the filter, nor its derivation. The reader is referred to Chapter 3 of Hamilton [19]. The Kalman filter is *stricto sensu*, not a machine learning method, as it relies on a parametric model. It consists of both a *state equation* and an *observation equation*, which must be linear.

For simplicity, we consider the case of one stock and one factor, but the method can be easily extended to the multivariate case.

24.4.1 ▪ First Case: Unobservable Betas

The hidden process is β. For simplicity, we choose the state equation an AR(1) model (see Chapter 12):

$$\beta_{t+1}(\omega) = c + \phi\beta_t(\omega) + \sigma v_{t+1}(\omega) \tag{24.10}$$

where v_{t+1} is IID Gaussian. Let $\xi_t = (\beta_t - c)/\phi$. The observation equation is

$$r_t(\omega) = -f_{t-1}\frac{c}{\phi} + f_{t-1}\xi_t(\omega) + u_t(\omega) \tag{24.11}$$

where both r_{t-1} and f_{t-1} are observable at time $t-1$. The Kalman filter will return $E[r_{t+1}|r_t]$ as well as $E[\beta_t|r_t]$, provided the parameters c, ϕ, and σ are known. If they are not known, the EM algorithm can be used (see Dempster, Laird, and Rubin [76]).

24.4.2 ▪ Second Case: Unobservable Market Price of Risk

We could model the market price of risk λ_t as a nonstationary hidden AR(1) process. The state equation is

$$\lambda_{t+1}(\omega) = c + \phi(f_t - \lambda_t(\omega)) + \sigma v_{t+1}(\omega) \tag{24.12}$$

You will work out the observation equation in the exercise section.

Further Reading

An excellent guide to understanding what factors are important in the US economy is the book by Ang [6]. The model of Kalman filtering with unobservable betas is borrowed from Hamilton [19]. Cochrane [21] gives more details about the cross-sectional regressions in the Fama–MacBeth method, as well as the GMM implementation. For a recent discussion of significance tests and data mining, read Giglio, Liao, and Xiu [77].

Exercises

24.1. (Medium) Refer to the data in equation (24.2). Define the portfolio excess return as $r^{e,p} = \frac{(r_1^e + r_2^e)}{2}$. Determine the beta of the portfolio β_p by the regression

$$r^{e,p}_{t+1} = \beta_p f_t + \text{ noise}$$

Is the beta of the portfolio average of the individual stocks β_1 and β_2? Discuss.

24.2. (Hard) Refer to section 24.3 on GMM. Find an expression for an estimate of the variance-covariance matrix of $\hat{\boldsymbol{\theta}}$. Take the case of a single security.

24.3. (Hard) Write the observation equation for the Kalman filter in the case where the market price of risk is unobservable and the state equation is (24.12). Betas are supposed known.

24.4. **Project** Implement a mixed clustering/Fama–MacBeth approach to determine the risk premia in different markets. The methodology is described in Patton and Weller [35].

24.5. **Project** Use Kalman filtering and the EM algorithm to determine beta using the state equation (24.10) and the observation equation (24.11).

24.6. **Project** Is it possible to do a joint estimation of the market price of risk and the betas by Kalman filtering? A simple start would be to take the logarithm of the product $\beta\lambda$ and consider the log of β and the log of λ to be Gaussian state variables. There is a subtle catch, however, when you take the logarithm, which you should work out.

24.7. (Medium) Why is GLS no longer the optimal estimate in the BARRA method?

Chapter 25

Estimation of Explicit Discount Factor Models

Explicit discount factor models have the following advantage over models of the return: it is easier to calibrate them so as to satisfy the no-arbitrage condition. Indeed, a positive SDF guarantees no arbitrage. Whether there is arbitrage or not in the economy (or in some sectors of the economy) is a separate question, but an asset manager with a view on this should apply explicit SDF methods.

In this chapter, we focus on estimation by GMM. An alternative would be estimation by OLS. In Chapter 8 we saw two types of moment conditions in GMM: we either set moments to zero, or, if that cannot be achieved, minimize their square. In this chapter, we will minimize their square.

We introduce a two-step procedure that is asymptotically efficient.[76] It is the nonlinear version of the Cochrane–Orcutt algorithm in GLS, which should yield enhanced efficiency of GLS over OLS.

25.1 ▪ Two-Stage GMM

Recall that $\mathbf{r}^e_{t+1}$ is the vector of excess returns of I securities at time $t+1$ and that M_{t+1} is the SDF for excess returns. The no-arbitrage relation (23.18) will hold approximately; that is, there will be some error $\mathbf{u}_{t+1}$, which we will try to minimize.[77] Writing explicitly the dependence of M_{t+1} on our model parameters $\boldsymbol{\theta}$, we have the following definitions.

Definition 25.1.

$$\mathbf{u}_{t+1}(\boldsymbol{\theta}) = M_{t+1}(\boldsymbol{\theta})\mathbf{r}^e_{t+1} \tag{25.1}$$

$$\mathbf{g}_T(\boldsymbol{\theta}) = E_T[\mathbf{u}(\theta)] \tag{25.2}$$

$$\mathbf{S}(\boldsymbol{\theta}) = \sum_{j=-\infty}^{\infty} E[\mathbf{u}_t(\theta)\mathbf{u}_{t-j}(\boldsymbol{\theta})']$$

It can be proved that the following estimator $\hat{\boldsymbol{\theta}}$ of $\boldsymbol{\theta}^*$ is consistent and asymptotically efficient:

$$\hat{\boldsymbol{\theta}} = \arg\min_{\boldsymbol{\theta}} \mathbf{g}_T(\boldsymbol{\theta})'\mathbf{S}^{-1}(\boldsymbol{\theta}^*)\mathbf{g}_T(\boldsymbol{\theta})$$

[76] Of course, it could be applied to other problems than finding the SDF.

[77] We can decompose this error into pricing error and uncertainty resolution (or market news). In the terminology of Chapter 2, uncertainty resolution is the intrinsic error, while pricing error is the sum of approximation and estimation errors.

Unfortunately, we have a circular problem: One needs to know the true value of $\boldsymbol{\theta}^*$ as well as the true distribution of the error terms to determine the optimal weighing matrix $\mathbf{S}$. Thus one adopts the two-stage GMM procedure. Start with any weighing matrix $\mathbf{W}$ (normally the identity), and then calculate the first-stage estimator $\hat{\boldsymbol{\theta}}_1$ and then the second-stage estimator $\hat{\boldsymbol{\theta}}_2$ according to the following algorithm.

ALGORITHM 25.1. Two-stage GMM with gradient descent.[78]

%Stage 1

1. `Initialize` $\mathbf{W} = \mathbf{I}, \boldsymbol{\theta}_0$
2. `Calculate` $\mathbf{g}_T(\boldsymbol{\theta}_0)$ `according to` (25.1), (25.2)
3. `Set` $\hat{\boldsymbol{\theta}}_1 = \arg\min_{\boldsymbol{\theta}} \mathbf{g}_T(\boldsymbol{\theta})'\mathbf{W}\mathbf{g}_T(\boldsymbol{\theta})$ `using` $\boldsymbol{\theta}_0$ `as initial value`

%Stage 2

4. `Calculate` $\mathbf{u}_t(\hat{\boldsymbol{\theta}}_1), \mathbf{g}_T(\hat{\boldsymbol{\theta}}_1)$ `according to` (25.1), (25.2)
5. `Set` $\hat{\mathbf{S}}(\hat{\boldsymbol{\theta}}_1) = \sum_{j=-J}^{J} E_T[\mathbf{u}_t(\hat{\boldsymbol{\theta}}_1)'\mathbf{u}_{t-j}(\hat{\boldsymbol{\theta}}_1)]$
6. `Set` $\hat{\boldsymbol{\theta}}_2 = \arg\min_{\boldsymbol{\theta}} \mathbf{g}_T(\boldsymbol{\theta})'\hat{\mathbf{S}}^{-1}(\hat{\boldsymbol{\theta}}_1)\mathbf{g}_T(\boldsymbol{\theta})$ `using` $\hat{\boldsymbol{\theta}}_1$ `as initial value`

Hansen's paper [20] shows that

$$var[\hat{\boldsymbol{\theta}}_2] = \frac{1}{T}(\mathbf{D}'(\hat{\boldsymbol{\theta}}_2)\mathbf{S}^{-1}(\boldsymbol{\theta}^*)\mathbf{D}(\hat{\boldsymbol{\theta}}_2))^{-1} \tag{25.3}$$

where $\mathbf{D}(\hat{\boldsymbol{\theta}}) = \frac{\partial \mathbf{g}_T(\hat{\boldsymbol{\theta}})}{\partial \hat{\boldsymbol{\theta}}}$. This estimator is a consequence of the delta method, as you will see in Exercise 25.2. Moreover, the statistic

$$T\mathbf{g}_T(\hat{\boldsymbol{\theta}}_2)'\mathbf{S}^{-1}(\boldsymbol{\theta}^*)\mathbf{g}_T(\hat{\boldsymbol{\theta}}_2)$$

has the χ^2 distribution with degree of freedoms equal to $I - K$, i.e., the number of moments minus the number of parameters.

Ergodicity

We saw in (23.19) the form of the SDF for excess returns imposed by the law of one price:

$$M_{t+1} = 1 - \tilde{\boldsymbol{\omega}}_t'\mathbf{r}_{t+1}^e$$

Thus calibrating the SDF means calibrating the SDF weights $\tilde{\omega}_t$. The next assumption we make is stationarity, i.e., that they are not direct functions of time:

$$\tilde{\boldsymbol{\omega}}_t' = \begin{bmatrix} \omega_1(\mathbf{x}_t;\boldsymbol{\theta}) \\ \vdots \\ \omega_I(\mathbf{x}_t;\boldsymbol{\theta}) \end{bmatrix}$$

where $\mathbf{x}$ is an N-vector of (hopefully) ergodic processes representing information available at time t, and $\omega_i : \mathbb{R}^N \times \mathbb{R}^K \to \mathbb{R}$ for $i = 1, \ldots, I$ are general functions to be calibrated. Once again, assuming ergodicity is key, so that

$$E_T[M_{t+1}\mathbf{r}_{t+1}^e] \simeq E[M_{t+1}\mathbf{r}_{t+1}^e]$$

[78] One can use either deterministic or stochastic gradient descent.

25.2 ▪ Interpreting the GMM procedure

25.2.1 ▪ First Stage

If we could, we would set $\hat{\boldsymbol{\theta}}_1$ such that the sample mean of the pricing error $\mathbf{g}_T(\hat{\boldsymbol{\theta}}_1) = \mathbf{0}$, but there are usually more moment conditions (returns times instruments) than there are free parameters. This is why we resort to minimization. Usually we set $\mathbf{W} = \mathbf{I}$. So the goal is to minimize a quadratic function of the sample mean of the pricing error. As you will be asked to show in Exercise 25.3, $\mathbf{g}_T$ is proportional to the vector $\boldsymbol{\alpha}$ from the CAPM. Note that the SDF weights may be nonlinear functions of $\boldsymbol{\theta}$, and thus nonlinear algorithms (such as neural networks) can be used.

25.2.2 ▪ Example of the First Stage

Let us drop the time subscript. There are $N = 2$ independent mean-zero "factors,"[79] x_1 and x_2, and $I = 2$ securities. The excess returns are denoted by r_1^e and r_2^e as usual.

Linearization

As we saw in Chapter 9, great care must be taken to choose good initial parameters for the nonlinear least squares algorithm. Otherwise, it may fail. We thus analyze a linearized version of the model and choose for initial conditions the parameters that solve the linearized model. To simplify matters, we suppose that $\mathbf{x}$ is mean-zero Gaussian, with $Var[x_1] = \sigma_1^2$ and $Var[x_2] = \sigma_2^2$. Regressing excess returns on factors, we find the following linearized relation:

$$\mathbf{r}^{e,lin}(\omega) = \bar{\mathbf{r}}^e + \mathbf{B}\mathbf{x}(\omega)$$

We choose the following parameters:

$$\bar{\mathbf{r}}^e = \begin{bmatrix} \bar{r}_2^e \\ \bar{r}_2^e \end{bmatrix} \quad \mathbf{B} = \begin{bmatrix} 1 & 0 \\ \rho & \sqrt{1-\rho^2} \end{bmatrix}$$

From (23.24), the SDF weights for this linearized model are thus $E[\mathbf{r}^{e,lin}\mathbf{r}^{e,lin\prime}]^{-1} * E[\mathbf{r}^{e,lin}]$. We choose a slightly simpler formula:

$$\begin{aligned} \tilde{\boldsymbol{\omega}}^{lin} &= \frac{1}{(\bar{r_1^e}^2 + \sigma_1^2)((r_2^e)^2 + \rho^2\sigma_1^2 + (1 - \rho^2)\sigma_2^2) - \rho^2\sigma_1^4)} \\ &\times \begin{bmatrix} (\bar{r}_2^e)^2 + \rho^2\sigma_1^2 + (1-\rho^2)\sigma_2^2 & -\rho\sigma_1^2 \\ -\rho\sigma_1^2 & (\bar{r}_1^e)^2 + \sigma_1^2 \end{bmatrix} \begin{bmatrix} \bar{r}_2^e \\ \bar{r}_2^e \end{bmatrix} \end{aligned}$$

Thus, one could imagine that a portfolio with the form

$$\tilde{\omega}_i = \tilde{\omega}_i^{lin} + h_i(\mathbf{r} - \bar{\mathbf{r}}^e - \mathbf{B}\mathbf{x}) \tag{25.4}$$

for some function h_i with $h_i(\mathbf{0}) = 0$ should be a good approximation.

Implementation as a Neural Network

In what follows, we choose $\sigma_1 = 1$, $\sigma_2 = 0.5$, $\rho = 0$, $\bar{r}_1^e = 0$, $\bar{r}_2^e = 0.125$. We choose an MLP with $L = 3$ layers, which outputs $\tilde{\omega}$. Some simple logic must then be added to output $M\mathbf{r}^e$, as Figure 25.1 shows.

[79]We henceforth stop using the notation f for factors, as this notation is heavily tilted in the empirical finance literature to represent "robust" factors in the economic sense.

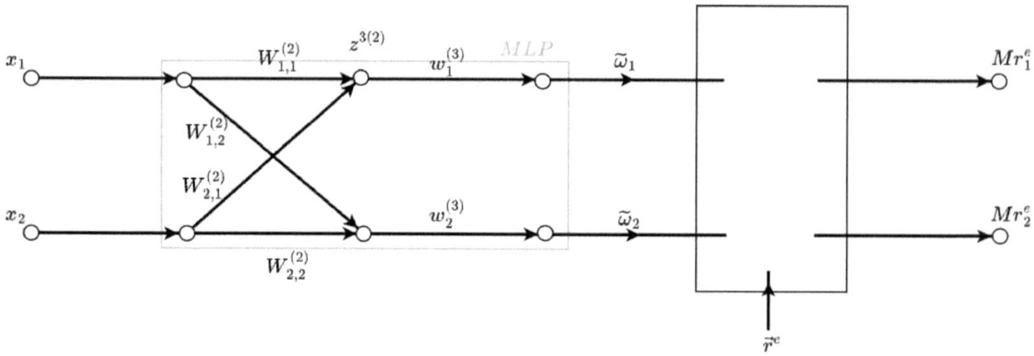

Figure 25.1: A three-layer MLP constructs the two functions $\tilde{\omega}_1$ and $\tilde{\omega}_2$, which are fed into some logic that outputs the errors $u_1 = Mr_1^e$ and $u_2 = Mr_2^e$ to be minimized either analytically or numerically. The network linking layers 2 and 3 is not complete. See a simple solution in the MATLAB code in Exercise 25.5 on how to implement an incomplete network.

For the hidden layer, the activation function is a sigmoid, and for the last layer, it is only the identity. To avoid having too many parameters, we chose a simplified structure for the network (see Chapter 11 for notation):

$$\begin{aligned} \mathbf{z}^{(2)} &= \mathbf{b}^{(2)} + \mathbf{W}^{(2)}\mathbf{x} \\ \mathbf{a}^{(2)} &= \boldsymbol{\sigma}^{(2)}(\mathbf{z}^{(2)}) \\ \begin{bmatrix} \omega_1(\mathbf{x},\boldsymbol{\theta}) \\ \omega_2(\mathbf{x},\boldsymbol{\theta}) \end{bmatrix} &= \begin{bmatrix} w_1^{(3)} a_1^{(2)} + b_1^{(3)} \\ w_2^{(3)} a_2^{(2)} + b_2^{(3)} \end{bmatrix} \end{aligned}$$

Since the network does not take the return as an input (note that it could have done so), we replace $\mathbf{r} - \bar{\mathbf{r}}^e$ by zero in (25.4); thus

$$\boldsymbol{\omega}(\mathbf{x},\boldsymbol{\theta}) \simeq \boldsymbol{\omega}^{lin} + \mathbf{h}(-\mathbf{B}\mathbf{x})$$

We chose for initial parameters of the third layer the identity, i.e., $\mathbf{w}^{(3)} = \mathbf{1}$ and $\mathbf{b}^{(3)} = \mathbf{0}$. Thus, the only possibility for h_i is to be a sigmoid, with slope $\frac{1}{4}$ when the argument is close to zero (since the third layer can modulate it). Thus, for small values of $\mathbf{x}$,

$$\boldsymbol{\omega}(\mathbf{x},\boldsymbol{\theta}) \simeq \tilde{\boldsymbol{\omega}}^{lin} - \frac{1}{4}\mathbf{B}\mathbf{x} \tag{25.5}$$

In order to avoid being trapped in a local minimum or diverging, the idea is to set the initial parameter $\boldsymbol{\theta}_0$ of the second layer so that the neural network resembles (25.5). In the linear regime, the output of the second layer is $\mathbf{a}^{(2)} \simeq \frac{\mathbf{z}^{(2)}}{4}$, so the output of the neural network, in first approximation, is

$$\boldsymbol{\omega}(\mathbf{x},\boldsymbol{\theta}_0) \simeq \frac{1}{4}\left(\mathbf{W}^{(2)}\mathbf{x} + \mathbf{b}^{(2)}\right)$$

Thus, if we take $\mathbf{W}^{(2)} = -\mathbf{B}$ and $\mathbf{b}^{(2)} = 4\tilde{\boldsymbol{\omega}}^{lin}$ as initial parameters, the neural network will start in the linear regime, as desired to avoid saturation.

The results are shown in Figure 25.2. Observe that the squared error is monotonously decreasing, as the problem is close to being convex. The MATLAB code is given in Exercise 25.5.

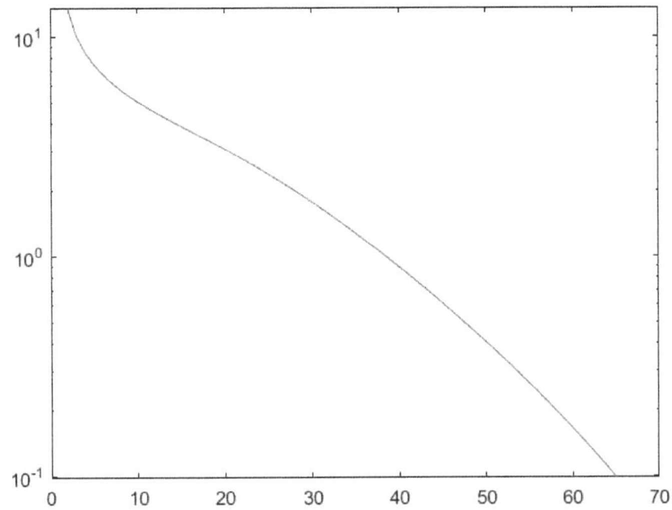

Figure 25.2: Plot of $\mathbf{g}'_T\mathbf{g}_T$ (squared error) with $T = 251$ as a function of the number of iterations. The data is $\mathbf{r^e}_t = \bar{\mathbf{r}}^{\mathbf{e}}_t + \mathbf{x}_t + 0.1 * (||\mathbf{x}_t||^2 - E[||\mathbf{x}_t||^2])$. The initial parameters are as in the text, with the exception of $\mathbf{b}^{(2)} = 0$.

25.2.3 ▪ Second Stage: Why $\hat{\mathrm{S}}^{-1}$

Some asset returns may have more pricing errors than others. For those assets the variance of the sample mean error $g_{T,i} = E_T(u_i)$ will be larger. The strategy will then be to pay less attention to pricing errors from assets with high variance of $u_{t+1,i}$. One could implement this idea by using a diagonal weight matrix composed of inverse variances of $E_T(u_i)$ on the diagonal. More generally, since asset returns are correlated, one might think of using the covariance matrix of $E_T(M_{t+1}\mathbf{r}^e_{t+1})$. For instance, if we have two assets, then

$$\hat{\mathbf{S}}(\hat{\boldsymbol{\theta}}_1) = \begin{bmatrix} \frac{1}{T}\sum_j u_{t+1,1}(\hat{\boldsymbol{\theta}}_1)u_{t+1-j,1}(\hat{\boldsymbol{\theta}}_1) & \frac{1}{T}\sum_j u_{t+1,1}(\hat{\boldsymbol{\theta}}_1)u_{t+1-j,2}(\hat{\boldsymbol{\theta}}_1) \\ \frac{1}{T}\sum_j u_{t+1,1}(\hat{\boldsymbol{\theta}}_1)u_{t+1-j,2}(\hat{\boldsymbol{\theta}}_1) & \frac{1}{T}\sum_j u_{t+1,2}(\hat{\boldsymbol{\theta}}_1)u_{t+1-j,2}(\hat{\boldsymbol{\theta}}_1) \end{bmatrix} \tag{25.6}$$

We do not advise populating the nondiagonal terms, though. In asset management, it is often costly to short assets. You will see in the exercises that long/short portfolio assets that are very correlated have strong (negative) off-diagonal coefficients of $\hat{\mathrm{S}}^{-1}$. Thus, the estimation will privilege long/short portfolios, which is not necessarily the goal. Hansen [20] shows that this estimator is asymptotically efficient, in a certain sense.

Exercises

25.1. (Medium) Show that when the model is linear, i.e., $m_{t+1}(\omega) = a + bx_{t+1}$ and $\boldsymbol{\theta} = (a, b)$, then two-stage GMM can be transformed into the Cochrane–Orcutt method of GLS. What is the disadvantage of this model?

25.2. (Easy) **The delta method**
Let g be a twice differentiable function and θ^* a constant such that $g(\theta^*) = 0$. Take $\hat{\theta}(\omega)$ to be a unbiased estimator of θ^*. Show that, in first approximation,

$$Var[\hat{\theta}] = \frac{Var[u]}{\frac{dg}{d\theta}}$$

25.3. (Easy) Show that $g_T = \frac{\alpha_T}{R_f}$, where $\alpha_T = E_T[\alpha]$ and α is the pricing error in the CAPM.

25.4. (Medium) Write the variance-covariance matrix of errors as

$$S = \begin{bmatrix} \sigma_1^2 & \rho\sigma_1\sigma_2 \\ \rho\sigma_1\sigma_2 & \sigma_2^2 \end{bmatrix}$$

Suppose ρ is close to 1. Discuss why two-stage GMM will result in better pricing of long/short portfolios than of long portfolios only. By "error," we mean the approximation/estimation error, not the intrinsic error. To make the problem simpler, you can assume that the latter is negligible.

25.5. The MATLAB code used to generate the results of the example of first-stage GMM is given below. Note that, to make it more readable, we did not vectorize the code. Add comments to each line of the code, relating it to equations in the text.

```
sigma2=0.5; re2bar=0.125; rho=0;
B=[1 0; rho sqrt(1-rho^2)];
rebar=[0; re2bar];
xmatrix=readmatrix("xmatrixGMM.csv");
rmatrix=readmatrix("RexcessGMM.csv");
BigT=length(rmatrix(:,1))
% Initialization
W2=-B; b2=[0;0]; W3=[1 0; 0 1]; b3=[0;0];
eta=0.1; Niter=100;
savecost=zeros(Niter,1);
counter=1; newcost=10;
while (counter<Niter) && (newcost>0.01)
    for t=1:BigT
     r=transpose(rmatrix(t,:)); x=transpose(xmatrix(t,:));
     a2=activatesigmoid(x,W2,b2);
     a3=W3*a2+b3;
     omega=a3;
     delta3=eye(2,2);
     delta2=a2.*(1-a2).*(transpose(W3)*delta3);
    dom1dW2(1,1)=delta2(1,1)*x(1); dom1dW2(1,2)=delta2(1,1)*x(2);
    dom1dW2(2,1)=delta2(1,2)*x(1); dom1dW2(2,2)=delta2(1,2)*x(2);
    dom1dW3(1,1)=delta3(1,1)*a2(1);
    dom1dW3(1,2)=0; %no branch, so W3[2,1]=0
    dom1dW3(2,1)=0; %no branch W3[2,1]=0
    dom1dW3(2,2)=0;
    dom1db2(1)=delta2(1,1); dom1db2(2)=delta2(1,2);
    dom1db3(1)=delta3(1,1); dom1db3(2)=0; %no branch
    dom2dW2(1,1)=delta2(2,1)*x(1); dom2dW2(1,2)=delta2(2,1)*x(2);
```

```
    dom2dW2(2,1)=delta2(2,2)*x(1); dom2dW2(2,2)=delta2(2,2)*x(2);
    dom2dW3(1,1)=0; dom2dW3(1,2)=0;
    dom2dW3(2,1)=0; dom2dW3(2,2)=delta3(2,2)*a2(2);
    dom2db2(1)=delta2(2,1); dom2db2(2)=delta2(2,2);
    dom2db3(1)=0; dom2db3(2)=delta3(2,2);
    Cost=(1-omega(1)*r(1)-omega(2)*r(2))^2*(r(1)^2+r(2)^2);
    dcostdom1=-r(1); dcostdom2=-r(2);
    dJdW2=Cost*(dcostdom1*dom1dW2+dcostdom2*dom2dW2);
    dJdb2=Cost*(dcostdom1*dom1db2+dcostdom2*dom2db2);
    dJdW3=Cost*(dcostdom1*dom1dW3+dcostdom2*dom2dW3);
    dJdb3=Cost*(dcostdom1*dom1db3+dcostdom2*dom2db3);
    W2=W2-eta*dJdW2/BigT; W3=W3-eta*dJdW3/BigT;
    b2=b2-eta*dJdb2/BigT; b3=b3-eta*dJdb3/BigT;
  end
  counter=counter+1;
  newcost=cost(W2,W3,b2,b3); savecost(counter)=newcost;
end
semilogy([1:counter], savecost(1:counter))

function costval=cost(W2,W3,b2,b3)
    costval=0;
     for i=t:BigT
        rt=transpose(rmatrix(t,:));
        xt=transpose(xmatrix(t,:));
         a2t=activatesigmoid(xtemp,W2,b2);
      omegat=W3*a2t+b3;
        MyCost=(1-omegat(1)*rt(1)-omegat(2)*rt(2))^2 *(rt(1)^2+rt(2)^2);
        costval=costval+MyCost;
    end
end
end
```

Chapter 26

Deep Learning in Asset Pricing

In this chapter, we apply the concepts developed in Chapters 8, 11, 12, 13, and 25 on deep feed-forward networks, long short-term memory (LSTM) networks, generative adversarial networks (GANs), and the generalized method of moments (GMM). We introduce three new concepts: conditional moment estimation, duality, and representation learning.

26.1 ▪ Conditional Moment Estimation

In Chapter 25 we focused on unconditional moment estimation, i.e., looking for a parameter $\boldsymbol{\theta}$ that minimizes the unconditional pricing error:

$$E[\mathbf{u}_t] \sim 0$$

We have so far not leveraged an important implication of our pricing models. If a pricing model holds true, then the unconditional pricing error should also be minimized; i.e., for each time t,

$$E_t[\mathbf{u}_{t+1}] \sim 0$$

Observe that this equation also holds true if for any stochastic process g

$$g_t E_t[\mathbf{u}_{t+1}] \sim 0$$

The tower law of expectations gives us

$$E[g_t \mathbf{u}_{t+1}] \sim 0$$

We can now boost our estimator by playing on the function g_t to extract more information from more meaningful error terms, i.e., error terms that have less variance. What should be the objective function? Using Hansen's notation (see Chapter 8), the GMM strategy is to minimize $(E_T[g(.)u(.+1)])^2$, while the OLS strategy is to minimize $E_T[(g(.)u(.+1))^2]$. It is clear that both benefit from ergodicity of the series $g_t u_{t+1}$. It can be argued that the GMM strategy is more efficient at making the series ergodic by directly working on setting the moments of the cross terms equal to zero and thus decreasing correlation.

26.2 ▪ Duality

Rather than solving least square problems, it is often more convenient to solve the dual problem. Hansen and Jagannathan pioneered this approach in asset pricing. We refer the reader to [80] for a full explanation of this approach. As explained in Chapter 4, the Lagrange dual of an optimization problem is a minimax problem. GANs are a tool of choice to solve minimax problems, as explained in Chapter 13.

26.3 ▪ A Representative Firm Model

In **representation learning**, one determines first an appropriate representation of the data. A fruitful representation of financial data is its decomposition into

- macroeconomic information (we will use the notation $\mathbf{h}_t$)
- firm-specific information (we will use the notation $\mathbf{k}_{i,t}$ for firm i).

This idea was used in the predictive factor model of Gu, Kelly, and Xiu [14], as well as that of Chen, Pelger, and Zhu [5].

Suppose that the goal is to calibrate the SDF weights $\tilde{\omega}$. In full generality, the SDF weight $\tilde{\omega}_i$ should be a function of all the information, that is,

$$\tilde{\omega}_i = \omega_i(\mathbf{h}_t, \mathbf{k}_{1,t}, ..., \mathbf{k}_{I,t}, \boldsymbol{\theta}_\omega)$$

In the model of Chen, Pelger, and Zhu [5], the function ω_i is the same for every firm i, which we could call the representative firm. We recognize the importance of firm-specific information by lumping it together with macro and idiosyncratic information into a vector $\mathbf{x}_{t,i}$:

$$\begin{aligned} \tilde{\omega}_i &= \omega(\mathbf{x}_{i,t}; \boldsymbol{\theta}_\omega) \\ \mathbf{x}_{i,t} &= (\mathbf{h}_t, \mathbf{k}_{i,t}) \end{aligned} \tag{26.1}$$

Thus for each sample t we apply the *same* function ω (determined by constant parameters $\boldsymbol{\theta}_\omega$) on contemporaneous data $\mathbf{x}_{i,t}$ for each firm. This idea leads to a drastic reduction in the number of parameters, as Figure 26.1 shows, and thus to a more efficient parameterization.

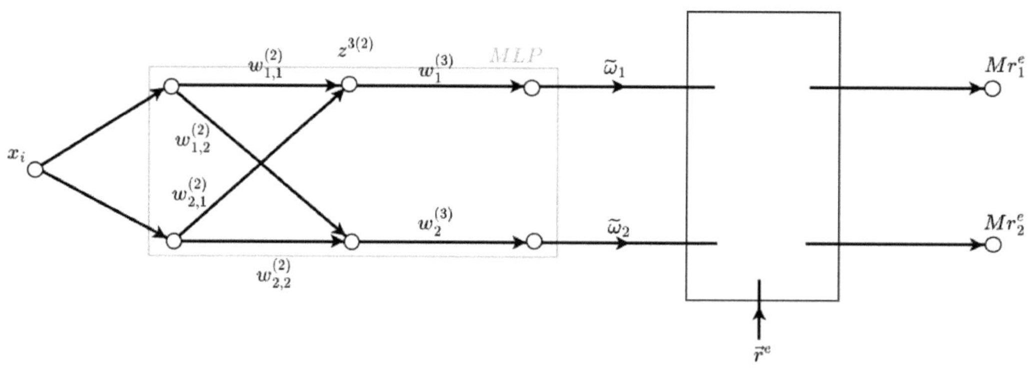

Figure 26.1: Compare with Figure 25.1. In the former, each firm was a different sample point. Here, we calibrate a representative firm. This is the reason why the input/output layer consists only of one node, the input data/SDF weight for the representative firm.

The same idea applies to models of the return. For instance, rather than choosing models of the form $E_t[r_{i,t+1}] = g_i(\mathbf{x}_{i,t})$, Gu, Kelly, and Xiu [14] describe models of the form $E_t[r_{i,t+1}] = g(\mathbf{x}_{i,t})$.

26.4 ▪ The CPZ Machine

Chen, Pelger, and Zhu (henceforth CPZ) were the first ones to implement *all* the concepts seen above in [5]:

- nonlinear estimation with regularized neural networks
- use of LSTM as a preprocessor for the MLP
- use of conditional estimation instead of unconditional estimation
- use of a GAN to find the optimum diagonal weight matrix $\mathbf{S}$ by solving the dual problem
- no-arbitrage (that is, an SDF-based model)
- stationary SDF weight function $\tilde{\omega}$
- representative firm modeling.

Rather than applying a single weight $\hat{S}_{i,i} = 1/g(\mathbf{x}_{i,t}; \boldsymbol{\theta}_\mathbf{g})$,[80] they use D different functions $g_d(\mathbf{x}_{i,t}; \theta_\mathbf{g})$ (called moment functions). The moment condition is then for each firm $n = 1, \ldots, I$ and moment condition $d = 1, \ldots, D$:

$$E_T[M r^e_n g_d] = 0 \tag{26.2}$$

We recall that the SDF for excess return is given in (23.19) by $M = 1 - \tilde{\omega}'\mathbf{r}^e$. We replace the root-finding problem (26.2) by the optimization problem:

$$\min_{\boldsymbol{\theta}_\omega} \max_{\boldsymbol{\theta}_\mathbf{g}} J(\boldsymbol{\theta}_\omega, \boldsymbol{\theta}_\mathbf{g})$$

where

$$J(\boldsymbol{\theta}_\omega, \boldsymbol{\theta}_\mathbf{g}) = \frac{1}{2} \sum_{n=1}^{I} \sum_{d=1}^{D} \left(\sum_{t=1}^{T} \left(1 - \sum_{i=1}^{I} \omega(\mathbf{x}_{i,t}; \boldsymbol{\theta}_\omega) r^e_{t+1,i} \right) r^e_{t+1,j} g_d(\mathbf{x}_{n,t}; \boldsymbol{\theta}_\mathbf{g}) \right)^2$$

That is, the portfolio weights are calculated at time t, following the conditional estimation method. The deep learning methodology is illustrated in Figure 26.2.

The architecture of the feedforward part of the SDF network is represented in Figure 26.3.

26.4.1 ▪ Results

Chen, Pelger, and Zhu [5] applied their machine to US stocks from CRSP, using monthly data from 1967 to 2016. This means that they use 46 time-varying firm characteristics and 178 macroeconomic time series. The dimensions of the problem are thus I = 32,000 (roughly), T = 600, and N = 224.

Figure 26.4 shows the importance of each macroeconomic variable. Figure 26.5 lists all firm-specific information that was used in the study. Figure 26.6 shows the relevance of each of them. Finally, Figure 26.7 shows the Sharpe ratio of the tangent portfolio. Recall that the Sharpe ratio is the expected return divided by the volatility of the investment. Its natural estimator is

$$\hat{SR} = \frac{E_T[F]}{\sqrt{Var_T[F]}}$$

[80] See Chapter 25 for the definition of the optimal weighting matrix $\hat{S}$ in GMM.

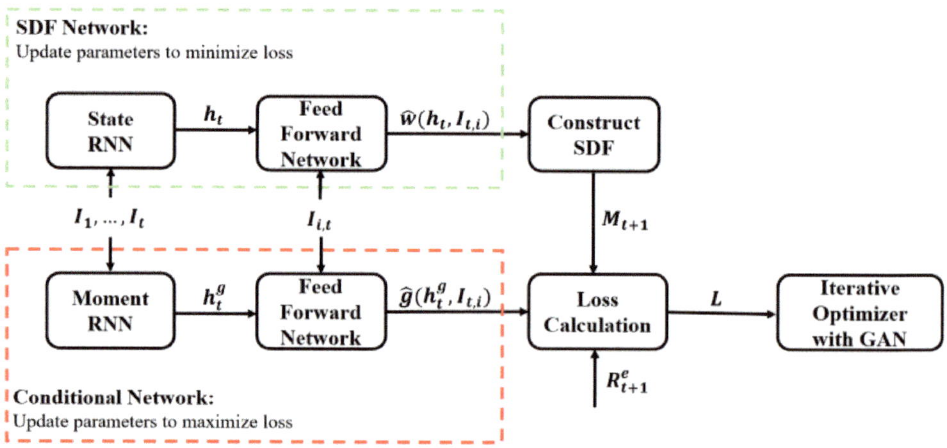

Figure 26.2: The deep learning architecture in the CPZ machine [5]; see definition of RNN in Chapter 12. Reprinted with permission from Chen, L., Pelger, M., and Zhu J., "Deep learning in asset pricing," Management Science, 0(0), (2023). Copyright © 2023 INFORMS.

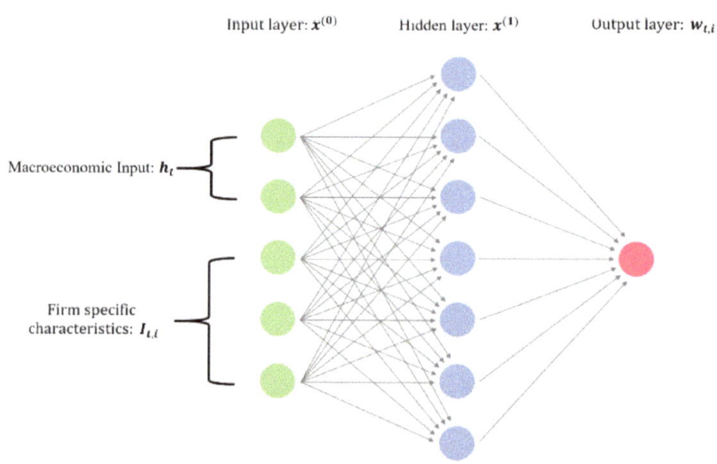

Figure 26.3: The feedforward subnetwork in the SDF network of the CPZ machine (Figure 2 in [5]). They use the notation $\mathbf{I}_{t,i}$ for firm-specific information, while we use the notation $\mathbf{k}_{i,t}$. Reprinted with permission from Chen, L., Pelger, M., and Zhu J., "Deep learning in asset pricing," Management Science, 0(0), (2023). Copyright © 2023 INFORMS.

The Sharpe ratio (in the test set) is equal to 75%. It is **20% higher than what is considered a "good" portfolio.**

26.4.2 ▪ Implementation of a Deep Learning GAN

We found it an important pedagogical exercise when we taught a class on this topic to focus on a particular implementation yet to reduce drastically the scope of the problem. In this section (complemented by exercises and projects), we focus on a simplified version of the CPZ

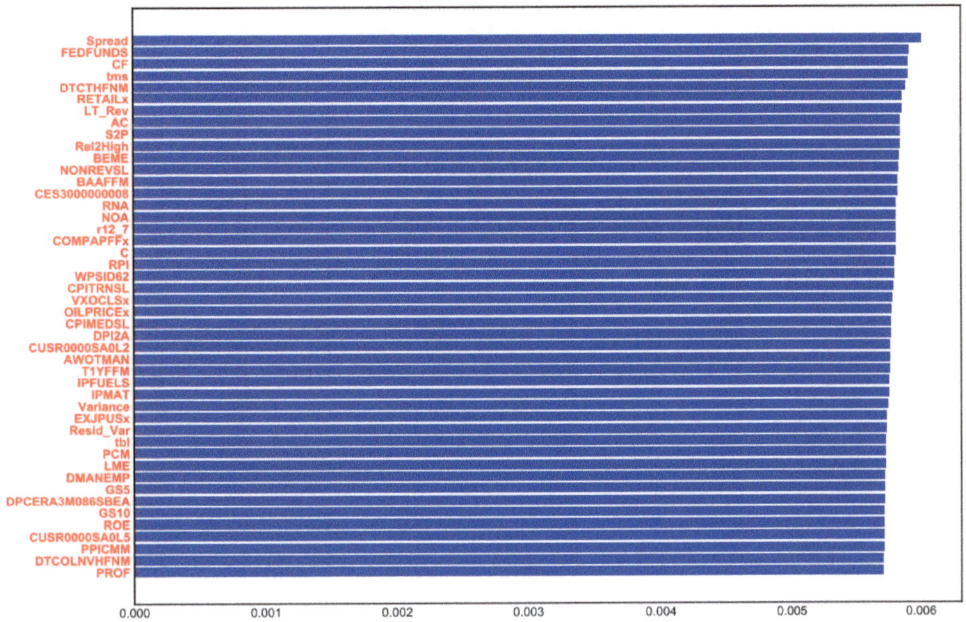

Figure 26.4: Importance of macroeconomic variables in the CPZ machine, Figure A.4 in [5]. Reprinted with permission from Chen, L., Pelger, M., and Zhu J., "Deep learning in asset pricing," Management Science, 0(0), (2023). Copyright © 2023 INFORMS.

	Variable	Description
	Past Returns	
(1)	r2_1	Short-term momentum
(2)	r12_2	Momentum
(3)	r12_7	Intermediate momentum
(4)	r36_13	Long-term momentum
(5)	ST_Rev	Short-term reversal
(6)	LT_Rev	Long-term reversal
	Investment	
(7)	Investment	Investment
(8)	NOA	Net operating assets
(9)	DPI2A	Change in property, plants, and equipment
(10)	NI	Net Share Issues
	Profitability	
(11)	PROF	Profitability
(12)	ATO	Net sales over lagged net operating assets
(13)	CTO	Capital turnover
(14)	FC2Y	Fixed costs to sales
(15)	OP	Operating profitability
(16)	PM	Profit margin
(17)	RNA	Return on net operating assets
(18)	ROA	Return on assets
(19)	ROE	Return on equity
(20)	SGA2S	Selling, general and administrative expenses to sales
(21)	D2A	Capital intensity
	Intangibles	
(22)	AC	Accrual
(23)	OA	Operating accruals
(24)	OL	Operating leverage
(25)	PCM	Price to cost margin
	Value	
(26)	A2ME	Assets to market cap
(27)	BEME	Book to Market Ratio
(28)	C	Ratio of cash and short-term investments to total assets
(29)	CF	Free Cash Flow to Book Value
(30)	CF2P	Cashflow to price
(31)	D2P	Dividend Yield
(32)	E2P	Earnings to price
(33)	Q	Tobin's Q
(34)	S2P	Sales to price
(35)	Lev	Leverage
	Trading Frictions	
(36)	AT	Total Assets
(37)	Beta	CAPM Beta
(38)	IdioVol	Idiosyncratic volatility
(39)	LME	Size
(40)	LTurnover	Turnover
(41)	MktBeta	Market Beta
(42)	Rel2High	Closeness to past year high
(43)	Resid_Var	Residual Variance
(44)	Spread	Bid-ask spread
(45)	SUV	Standard unexplained volume
(46)	Variance	Variance

Figure 26.5: List of firm idiosyncratic information variables in the CPZ machine (Table A.II in [5]). Reprinted with permission from Chen, L., Pelger, M., and Zhu J., "Deep learning in asset pricing," Management Science, 0(0), (2023). Copyright © 2023 INFORMS.

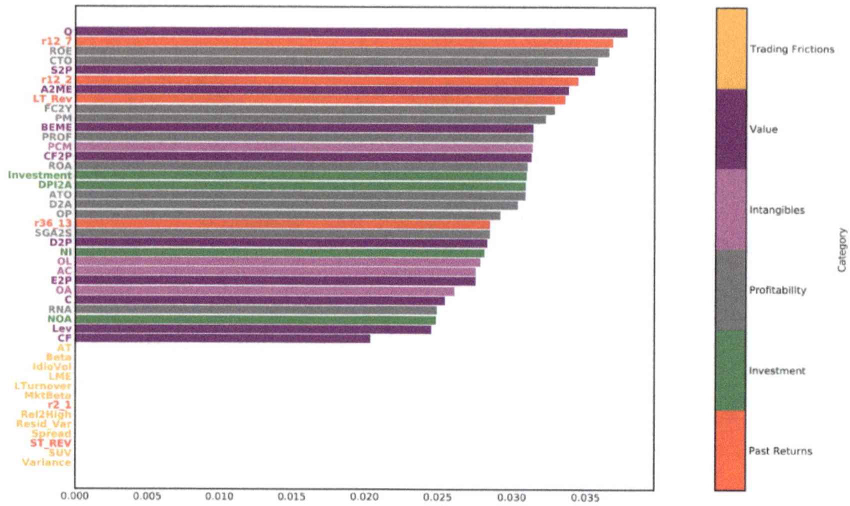

Figure 26.6: Relative importance of a firm characteristic information in the CPZ machine (Figure A.8 in [5]). Reprinted with permission from Chen, L., Pelger, M., and Zhu J., "Deep learning in asset pricing," Management Science, 0(0), (2023). Copyright © 2023 INFORMS.

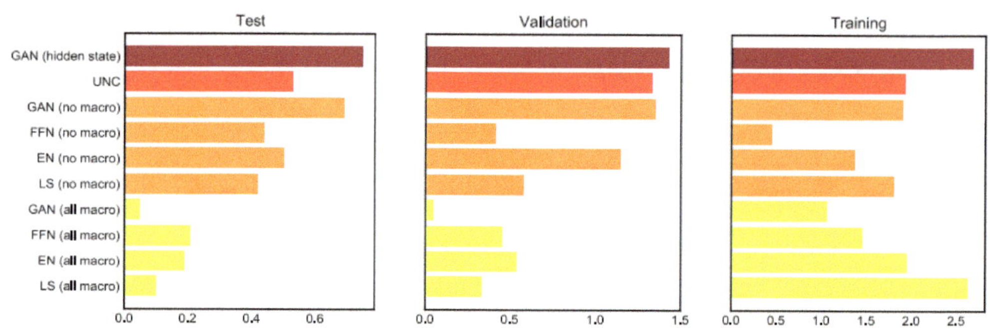

Figure 26.7: The reward for reading 200 pages of text: the Sharpe ratio in the CPZ machine is 20% better than what is considered now "a good investment" (Table 1 in [5]). Reprinted with permission from Chen, L., Pelger, M., and Zhu J., "Deep learning in asset pricing," Management Science, 0(0), (2023). Copyright © 2023 INFORMS.

machine seen above. In particular, we do not use preprocessors (but see Exercise 26.7). Our algorithmic choices were made for simplicity, and we do not pretend they reflect the exact CPZ architecture.

We will use two three-layer networks to represent ω and $\mathbf{g}$. In our example, we will take the dimension of $\mathbf{h}_t$ to be one (market return) and of $\mathbf{k}_{i,t}$ to be one also (weather). Thus the dimension of $\mathbf{x}$ will be $N = 2$. We will take $D = 2$ moment conditions. For simplicity, we omit superscripts on the activation function. Then

$$\omega(\mathbf{x};\boldsymbol{\theta}_\omega) = \boldsymbol{\sigma}(\mathbf{w}_\omega^{(3)}\boldsymbol{\sigma}(\mathbf{W}_\omega^{(2)}\mathbf{x} + \mathbf{b}_\omega^{(2)}) + b_\omega^{(3)})$$
$$\mathbf{g}(\mathbf{x};\boldsymbol{\theta}_\mathbf{g}) = \boldsymbol{\sigma}(\mathbf{W}_\mathbf{g}^{(3)}\boldsymbol{\sigma}(\mathbf{W}_\mathbf{g}^{(2)}\mathbf{x} + \mathbf{b}_\mathbf{g}^{(2)}) + \mathbf{b}_\mathbf{g}^{(3)})$$

We then have nine parameters for $\boldsymbol{\theta}_\omega$ and 12 parameters for $\boldsymbol{\theta}_g$. This gives a total of $K = 21$ parameters:

$$\boldsymbol{\theta}_\omega = \begin{bmatrix} \mathbf{W}_\omega^{(2)}(1,1) \\ \mathbf{W}_\omega^{(2)}(1,2) \\ \mathbf{W}_\omega^{(2)}(2,1) \\ \mathbf{W}_\omega^{(2)}(2,2) \\ \mathbf{b}_\omega^{(2)}(1) \\ \mathbf{b}_\omega^{(2)}(2) \\ \mathbf{w}_\omega^{(3)}(1) \\ \mathbf{w}_\omega^{(3)}(2) \\ b_\omega^{(3)} \end{bmatrix} \qquad \boldsymbol{\theta}_\mathbf{g} = \begin{bmatrix} \mathbf{W}_g^{(2)}(1,1) \\ \mathbf{W}_g^{(2)}(1,2) \\ \mathbf{W}_g^{(2)}(2,1) \\ \mathbf{W}_g^{(2)}(2,2) \\ \mathbf{b}_g^{(2)}(1) \\ \mathbf{b}_g^{(2)}(2) \\ \mathbf{W}_g^{(3)}(1,1) \\ \mathbf{W}_g^{(3)}(1,2) \\ \mathbf{W}_g^{(3)}(2,1) \\ \mathbf{W}_g^{(3)}(2,2) \\ \mathbf{b}_g^{(3)}(1) \\ \mathbf{b}_g^{(3)}(2) \end{bmatrix}$$

We choose $T = 10$ samples and $I = 5$ (five firms). For notational convenience in coding, we split J to give

$$J = \frac{1}{2}\sum_{n=1}^{I}\sum_{d=1}^{D} J(n,d)^2 \tag{26.3}$$

$$J(n,d) = \sum_{t=1}^{T}\left(1 - \sum_{i=1}^{N}\omega(\mathbf{x}_{i,t},\boldsymbol{\theta}_\omega) r_{i,t+1}^e\right) r_{j,t+1}^e g_d(\mathbf{x}_{n,t};\boldsymbol{\theta}_\mathbf{g}) \tag{26.4}$$

In both the minimization and maximization parts, the stochastic gradient algorithm will pick a random security (n) and a random moment condition (d) so as to optimize only $J(n,d)^2$.

ALGORITHM 26.1. Stochastic gradient algorithm for GANs.

Set learning rate η, NbIter
Set initial value for $\boldsymbol{\theta}_\omega^0, \boldsymbol{\theta}_\mathbf{g}^0$
For iter =1 to NbIter
  Minimization
  1. Set $k = 0$
    For each j and t, calculate $\mathbf{g}(\mathbf{x}_{j,t};\boldsymbol{\theta}_\mathbf{g}^0)$
  2. Repeat
    For each i and t, calculate $\omega(\mathbf{x}_{i,t};\boldsymbol{\theta}_\omega^k)$
    Choose at random an integer n from $\{1,\dots,I\}$
      and d from $\{1,\dots,D\}$
    Set $\boldsymbol{\theta}_\omega^{k+1} = \theta_\omega^k - \eta\nabla_{\boldsymbol{\theta}_\omega} J(n,d)^2$
    $k \leftarrow k+1$
  3. Until k =Nbiter
    Set $\boldsymbol{\theta}_\omega^0 = \boldsymbol{\theta}_\omega^{Niter+1}$
  Maximization
  4. Set $k = 0$
    For each i and t calculate $\omega(\mathbf{x}_{i,t};\boldsymbol{\theta}_\omega^0)$

```
5.  Repeat
       For each i and t, calculate g(x_{i,t}; θ_g^k)
       Choose at random an integer n from {1,...,I}
          and d from {1,...,D}
       Set θ_{g_d}^{k+1} = θ_{g_d}^k + η∇_{θ_{g_d}} J(n,d)^2
       k ← k+1
6.  Until k =Nbiter
    Set θ_g^0 = θ_g^{Nbiter+1}
end
```

Exercises

26.1. (Easy) Recall the notation (26.4). Show that

$$\nabla_{\boldsymbol{\theta}_\omega} J(n,d)^2 = -2J(n,d)\sum_{t=1}^{T} r^e_{t+1,n} g_d(\mathbf{x}_{n,t};\boldsymbol{\theta}_{\mathbf{g}})\sum_{i=1}^{N}\nabla_{\boldsymbol{\theta}_\omega}\omega(\mathbf{x}_{i,t};\boldsymbol{\theta}_\omega) r^e_{t+1,i}$$

26.2. (Medium) In our implementation,

$$\nabla_{\boldsymbol{\theta}_\omega}\omega(\mathbf{x}^{t,\imath};\boldsymbol{\theta}_\omega) = \nabla_{\boldsymbol{\theta}_\omega}\boldsymbol{\sigma}(\mathbf{w}^{(3)}_\omega \boldsymbol{\sigma}(\mathbf{W}^{(2)}_\omega \mathbf{x}_{i,t}+\mathbf{b}^{(2)}_\omega) + b^{(3)}_\omega) \tag{26.5}$$

Give explicit formulae for this gradient, using the same notation as in Chapter 11:

$$\delta^{(3)}_\omega \equiv \frac{\partial \omega}{\partial z^{(3)}_\omega} = \sigma'(z^{(3)}_\omega)$$

$$\boldsymbol{\delta}^{(2)}_\omega \equiv \frac{\partial \omega}{\partial \mathbf{z}^{(2)}_\omega} = \boldsymbol{\sigma}'(\mathbf{z}^{(2)}_\omega)\circ \mathbf{w}^{(3)}_\omega \delta^{(3)}_\omega$$

26.3. (Medium) Calculate analytically $\nabla_{\boldsymbol{\theta}_{\mathbf{g}_d}} J(n,d)^2$, like in Exercises 26.1 and 26.2.

26.4. Code in MATLAB the stochastic gradient algorithm for GANs and apply to the data found in Section 25.2.2.

26.5. Suppose a linear form for ω, that is,

$$\omega(x) = a + bx \tag{26.6}$$

Solve analytically for a and b.

26.6. **Project 1**
The algorithm coded in Exercise 26.4 will probably yield disappointing results, depending on your choice of hyperparameters. If so, you should try to make the MLP work first, by taking out the GAN. That is, set $D = 1$ and $g_D = 1$. Change the learning rate, and if it does not work, replace the stochastic gradient algorithm with the gradient algorithm with line search. Add a regularizer on the parameters if needed.

26.7. **Project 2**
Recall Exercise 12.2, where you coded a simple LSTM. Add this LSTM as a preprocessor to the MLP used for the SDF in the stochastic gradient algorithm for GANs.

26.8. **Project 3**

Take $I = 2$ securities: a stock, and a call option on that stock. Download series on the return of these instruments. These series will be highly correlated. The information (x) will be the stock return.

(a) Add a regularizer that will increase the likelihood for the SDF to be positive. (The no-arbitrage constraint is universally used in option pricing.)

(b) Take a linear form of the SDF weight, like in (26.6), that is

$$\omega(x) = \begin{bmatrix} a_1 + b_1 x \\ a_2 + b_2 x \end{bmatrix}$$

(c) Choose a nonlinear function for the option weight, that is, modify $\omega_2(x)$, and try to fit the model.

Hint: For those of you who know derivatives, reuse the formula for the delta and gamma of the call option as initial parameters.

Chapter 27

Empirical Results

One question the reader might ask after 26 chapters is, how good are the methods that were described in this book, in terms of selecting an optimal portfolio? The goal of this chapter is to try to answer this question.

27.1 ▪ Dataset

We are using four indices. To keep in mind, we want a dataset that is easily accessible on the internet as well as covers a wide range of stocks. The datasets we are using are

- S&P 500 (SPY)
- MSCI Europe, Australasia, and Far East (EFA)
- iShares 20+ Year Treasury Bond ETF (TLT)
- iShares Investment Grade Corporate Bond ETF (LQD).

Also, for the risk-free rate, we are using the one-month US Treasury rate. The data range is from 2002 to 2022, and we are using monthly data.

27.2 ▪ Methodology

We also introduced rebalancing schemes for traditional asset allocation techniques. We used five years of data $(t - 60)$ to calculate today's (t) weight allocation and held the portfolio for one month.

27.3 ▪ Key Measures

We focus on two measures:

- the alpha of the "optimal" portfolio
- the Sharpe ratio of the "optimal" portfolio.

	Monthly net return (%)	Volatility of monthly returns (%)	Monthly Sharpe ratio
Mean-variance weights (tangent portfolio)	0.563	3.75	0.12
Equal weights	0.483	2.98	0.128
Minimum variance	0.329	2.845	0.08

Table 27.1: Our results based on selected portfolios for the period 2007-09-01 to 2023-05-01. A rule of thumb, to annualize volatility, is to multiply monthly volatility by $\sqrt{12}$.

	Raw return	Volatility	Sharpe ratio
Mean-variance weights (tangent portfolio)	6.06	11.59	0.07
Equal weights	10.00	8.66	0.54
Minimum variance	7.96	5.12	0.52

Table 27.2: Results for the period 1973 to 2011; numbers reported from [6].

27.4 ▪ Traditional Asset Allocation Techniques

We study the performance of the following traditional asset allocation techniques:

- mean-variance weights (tangent portfolio)
- equal weights
- minimum variance.

The results are shown in Table 27.1. For reference, we chose to compare them with Ang's results [6]; see Table 27.2. In Ang's table, he choose a much longer period, from 1973 to 2011, with the same portfolio exposure with US stocks, US corporate bonds, US government bonds, and international stocks. Portfolios fluctuate wildly from month to month in the methods that require inversion of the variance-covariance matrix. In order to limit such fluctuation, we constrained each weight to be between -1 and 2. For more sophisticated methods, please consult Chapter 18.

27.5 ▪ Linear Factor Modeling

In this section, we ran the same four datasets with Fama–French three-factor and Fama–French five-factor models to compare the alphas of each dataset. The results are shown in different time periods including 6 months, 2 years, 5 years, 10 years, and 20 years. Tian [78] made a systematic study and concluded, by systematic data mining, that she could find better factors out of sample. Nevertheless, we chose the Fama–French factors due to their wide acceptance in academia. Recall the Fama–French model:

- Mkt-RF (risk-free rate) is the market risk premium
- SMB (small minus big) is the excess returns of small-cap stocks over large cap stocks
- HML (high minus low) is the excess returns of stocks with high book-to-market ratio over low book-to-market ratio (value stock over growth stock)

Period	Factor	EFA	LQD	SPY	TLT
6m	Mkt-RF	0.011200	0.006100	0.010500	0.005900
	SMB	−0.023100	−0.014800	−0.001700	−0.020100
	HML	0.000700	−0.003200	0.000600	−0.005900
	Alpha	0.015100	0.001333	0.001884	−0.010965
2y	Mkt-RF	0.008700	0.004000	0.010300	0.003200
	SMB	−0.002600	−0.001800	−0.001800	−0.004600
	HML	0.003200	−0.000700	0.000700	−0.003500
	Alpha	−0.007595	−0.007305	0.001864	−0.009709
5y	Mkt-RF	0.008400	0.003100	0.010100	0.000800
	SMB	−0.001400	−0.000800	−0.001900	−0.003200
	HML	0.002600	−0.001100	0.000700	−0.004900
	Alpha	−0.003584	−0.001939	0.001275	−0.002511
10y	Mkt-RF	0.008800	0.002600	0.010100	0.000100
	SMB	−0.002400	−0.001000	−0.002000	−0.002800
	HML	0.001600	−0.001200	0.000600	−0.005000
	Alpha	−0.004311	−0.000970	0.000532	0.000637
20y	Mkt-RF	0.010200	0.002200	0.010100	−0.000800
	SMB	−0.002400	−0.001200	−0.001900	−0.002800
	HML	0.001900	−0.000900	0.000500	−0.002900
	Alpha	−0.001360	0.002002	0.000889	0.005080

Table 27.3: Portfolio exposure to Fama–French three-factor model.

- RMW (robust minus weak) is the excess returns with companies with high operating profitability over low operating profitability

- CMA (conservative minus aggressive) is the excess returns with companies with low investment over high investment.

The results are shown in Table 27.3 and Table 27.4. From these two tables, we can make the following observations:

- **Alphas:** For both the three-factor and five-factor models, we can see that the intercepts varied for all exchange-traded funds (ETFs) based on different time periods. We can conclude that the ETFs had different excess returns that couldn't be explained by the exposure factors of the models.

- All the ETFs had a positive exposure to the Mkt-RF factor. It means that these four ETFs tend to perform well when the market performs well.

- The SMB exposure is mostly negative, meaning that these ETFs are more exposed to larger companies. This is not a surprising result because these four ETFs are focused on large and developed markets.

- We can see that the SPY had a consistent overperformance across all time periods.

From this experiment, we see that running very simple models allows us to gain a very rich understanding of these ETFs.

Period	Factor	EFA	LQD	SPY	TLT
6m	Mkt-RF	0.009500	0.005100	0.010200	0.007900
	SMB	−0.006900	−0.003700	−0.000100	−0.014800
	HML	−0.012900	−0.012900	−0.000300	−0.021300
	RMW	0.007400	0.004900	0.000900	−0.001300
	CMA	0.010400	0.007500	0.000600	0.014300
	Alpha	0.010692	−0.001672	0.001420	−0.011720
2y	Mkt-RF	0.008600	0.003700	0.010200	0.002900
	SMB	−0.001400	−0.000600	−0.001100	−0.003000
	HML	0.002300	−0.001200	0.000400	−0.003800
	RMW	0.001700	0.001900	0.001000	0.002500
	CMA	0.001200	0.000300	0.000100	−0.000200
	Alpha	−0.009527	−0.009037	0.000899	−0.011791
5y	Mkt-RF	0.008500	0.002900	0.010000	0.000700
	SMB	−0.000900	−0.000500	−0.001600	−0.002900
	HML	0.001900	−0.001000	0.000700	−0.005100
	RMW	0.000400	0.001000	0.000600	0.000700
	CMA	0.001400	−0.000700	−0.000200	0.000300
	Alpha	−0.004412	−0.001963	0.001151	−0.002909
10y	Mkt-RF	0.008900	0.002400	0.010000	−0.000100
	SMB	−0.002400	−0.000400	−0.001800	−0.001900
	HML	0.001400	−0.001300	0.000500	−0.005800
	RMW	−0.000000	0.001500	0.000500	0.002100
	CMA	0.000500	−0.000400	−0.000000	0.001100
	Alpha	−0.004408	−0.001138	0.000448	0.000050
20y	Mkt-RF	0.010000	0.002200	0.010100	−0.000700
	SMB	−0.002900	−0.001000	−0.001700	−0.002500
	HML	0.002200	−0.000800	0.000300	−0.003200
	RMW	−0.001700	0.000700	0.000400	0.001000
	CMA	−0.000600	−0.000500	0.000300	0.000400
	Alpha	−0.000665	0.001849	0.000697	0.004664

Table 27.4: Portfolio exposure to Fama–French five-factor model.

27.6 ▪ Deep Learning in Asset Pricing

Moving away from the linear factor models, we now investigate more complicated nonlinear models. In this section, we will be discussing the results from the three papers that we think are the frontiers of the topics of deep learning in asset pricing. Table 27.5 shows the results from the three papers mentioned below with methods described in the papers.

- "Empirical Asset Pricing via Machine Learning" by Gu, Kelly, and Xiu [14]:

  OLS-3 Ordinary least squares with size, book-to-market ratio, and 12-month momentum

  OLS-7 OLS-3 plus absolute accruals, return on asset, asset growth, and growth in common shareholder equity

  OLS-15 OLS-7 plus dividend to price, mom36m (36-month momentum), beta, return volatility, share turnover, leverage, sales to price

Paper	Year	Test period	Method/model	Sharpe ratio	R^2
Gu, Kelly, and Xiu	2020	1957 to 2016	OLS-3	0.61	0.16
			OLS-7	0.74	0.74
			OLS-15	0.86	0.19
			RF	0.98	1.48
			NN3	1.20	0.98
Chen, Pelger, and Zhu	2019	1967 to 2016	GAN	0.75	0.23
Jensen et al.	2022	1952 to 2020	Portfolio-ML	1.43	N/A
			Multiperiod-ML	0.95	N/A
			Static-ML	1.06	N/A
			Portfolio sort	1.10	N/A
			Markowitz-ML	2.00	N/A
			Two tuning layers		
			Multiperiod-ML*	1.33	N/A
			Static-ML*	1.36	N/A

Table 27.5: Comprehensive summary results.

Random forest

NN3 Neural network with three hidden layers with 32, 16, and 8 neurons.

- "Deep Learning in Asset Pricing" by Chen, Pelger, and Zhu [5]:

  GAN Generative adversarial network.

- "Machine Learning and the Implementable Efficient Frontier" by Jensen et al. [79]:

  Portfolio-ML Machine learning directly about portfolio weights

  Multiperiod-ML Machine learning about expected returns across horizons

  Static-ML Static trading cost optimization method

  Portfolio sort Long/short the 10% of stocks with the highest/lowest 1-month expected return

  Markowitz-ML Optimal portfolio absent trading cost

  One tuning layers vs. two tuning layers The authors perform a hyperparameter search using ridge regression on RF transformed features for methods with one tuning layer. For Multiperiod-ML $*$ and Static-ML$*$, the authors add a second tuning layer to adjust inputs for expected return, covariance, and trading cost.

27.7 ▪ To Probe Further

This is a rapidly moving area. The paper [14] is cited by many authors, and we expect it will continue to be so. A good way to keep up with the material is to read papers that quote [14]. Few practitioners publish their results in an amenable way. An exception is [81].

Bibliography

[1] MERTON, ROBERT C., *Theory of rational option pricing*, The Bell Journal of Economics and Management Science, 4(1), 141–183, (1973), https://doi.org/10.2307/3003143. (Cited on p. 6)

[2] BRENNAN, M., SCHWARTZ, E., AND LAGNADO, R., *Strategic asset allocation*, Journal of Economic Dynamics and Control, 21, 1377–1403, (1997). (Cited on p. 6)

[3] LOPEZ DE PRADO, M., *Advances in Financial Machine Learning*, John Wiley and Sons, New York, 2018. (Cited on p. 8)

[4] ROSS, S. A., *The arbitrage theory of capital asset pricing*, Journal of Economic Theory, 13(3), 341–360, (1976), https://doi.org/10.1016/0022-0531(76)90046-6. (Cited on p. xviii)

[5] CHEN, L., PELGER, M., AND ZHU, J., *Deep Learning in Asset Pricing*, preprint, arXiv:1904.00745[q-fin.ST], (2019), https://doi.org/10.48550/arXiv.1904.00745. (Cited on pp. 5, 115, 214, 215, 216, 217, 218, 227)

[6] ANG, A., *Asset Management: A Systematic Approach to Factor Investing*, Financial Management Association Survey and Synthesis Series, Oxford University Press, New York, 2014, https://doi.org/10.1093/acprof:oso/9780199959327.001.0001. (Cited on pp. 8, 135, 137, 203, 224)

[7] MITCHELL, T. M., *Machine Learning*, McGraw-Hill, New York, 1997. (Cited on p. 13)

[8] GOODFELLOW, I., BENGIO, Y., AND COURVILLE, A., *Deep Learning*, MIT Press, Cambridge, MA, 2016. (Cited on pp. 13, 17, 102, 106, 114)

[9] FEDUS, W., ZOPH, B., AND SHAZEER, N., *Switch transformers: Scaling to trillion parameter models with simple and efficient sparsity*, The Journal of Machine Learning Research, 23(1), 5232–5270, (2022). (Cited on p. 17)

[10] PAPYAN, V., HAN, X. Y., AND DONOHO, D. L., *Prevalence of neural collapse during the terminal phase of deep learning training*, Proceedings of the National Academy of Sciences, 117(40), 24652–24663, (2020), https://doi.org/10.1073/pnas.2015509117. (Cited on p. 17)

[11] BELKIN, M., HSU, D., MA, S., AND MANDAL, S., *Reconciling modern machine-learning practice and the classical bias–variance trade-off*, Proceedings of the National Academy of Sciences, 116(32), 15849–15854, (2019), https://doi.org/10.1073/pnas.1903070116. (Cited on pp. 17, 79, 103)

[12] WOLPERT, D. H, *The lack of a priori distinction between learning algorithms*, Neural Computations, **8**(7), 1341–1390, (1996). (Cited on p. 17)

[13] NAGEL, S., *Machine Learning in Asset Pricing*, Princeton University Press, Princeton, NJ, 2021. (Cited on p. 18)

[14] GU, S., KELLY, B., AND XIU, D., *Empirical asset pricing via machine learning*, The Review of Financial Studies, 33(5), 2223–2273, (2020), https://doi.org/10.1093/rfs/hhaa009. (Cited on pp. 19, 84, 214, 226, 227)

[15] GILL, P., MURRAY, W., AND M. WRIGHT, *Practical Optimization*, Academic Press, New York, 1982. (Cited on p. 79)

[16] BAZARAA, M., AND SHETTY, C. M., *Nonlinear Programming, Theory and Algorithms*, 3rd Edition, John Wiley and Sons, New York, 2006. (Cited on pp. 29, 37, 41, 79, 117)

[17] INTRILIGATOR, M., *Mathematical Optimization and Economic Theory*, SIAM, Philadelphia, 1971. (Cited on pp. 41, 127)

[18] KORN, R., *Optimal Portfolios*, World Scientific, River Edge, NJ, 1997, https://doi.org/10.1142/3548. (Cited on pp. 41, 168)

[19] HAMILTON, J., *Time Series Analysis*, Princeton University Press, Princeton, NJ, 1994. (Cited on pp. 64, 68, 71, 111, 202, 203)

[20] HANSEN, L. P., *Large sample properties of generalized method of moments estimators*, Econometrica, 50(4), 1029–1054, (1982), https://doi.org/10.2307/1912775. (Cited on pp. 68, 71, 206, 209)

[21] COCHRANE, J., *Asset Pricing*, Princeton University Press, Princeton, NJ, 2005. (Cited on pp. 71, 134, 178, 203)

[22] SILVERMAN, B. W., *Density Estimation for Statistics and Data Analysis*, Chapman and Hall, London, 1986, https://dx.doi.org/10.1007/978-1-4899-3324-9. (Cited on p. 84)

[23] SKLAR, A., *Fonctions de répartition à n dimensions et leurs marges*, Publications de l'Institut de statistique de l'Université de Paris, 8, 229–231, (1959). (Cited on p. 85)

[24] HASTIE, T., TIBSHIRANI, R., AND FRIEDMAN, J. H., *The Elements of Statistical Learning: Data Mining, Inference, and Prediction*, 2nd Edition, Springer, New York, 2009. (Cited on pp. 85, 89, 90)

[25] TSAY, R., *Analysis of Financial Time Series.* Wiley, (2005). (Cited on pp. 86, 111)

[26] PHOA, W., *Empirical yield-curve dynamics and yield-curve exposure*, in The Handbook of Fixed Income Securities, Ninth Edition, McGraw-Hill Education, New York, 2021, Chapter 33. (Cited on p. 87)

[27] LLOYD, S., *Least squares quantization in PCM*, IEEE Transactions on Information Theory, 28(2), 129–137, (1982), https://doi.org/10.1109/TIT.1982.1056489. (Cited on p. 90)

[28] FORGY, E. W., *Cluster analysis of multivariate data: Efficiency vs interpretability of classifications*, Biometrics, 21, 768–780, (1965). (Cited on p. 90)

[29] FISHER, R. A., *The use of multiple measurements in taxonomic problems*, Annals of Eugenics, 7(2), 179–188, (1936), https://doi.org/10.1111/j.1469-1809.1936.tb02137.x. (Cited on p. 89)

[30] TANG, G., TIAN, R., AND WU, B., *An overview of clustering methods in the financial world*, in Proceedings of the 2022 7th International Conference on Financial Innovation and Economic Development (ICFIED 2022), Atlantis Press, Dordrecht, 2022, 524–529, https://doi.org/10.2991/aebmr.k.220307.084. (Cited on p. 91)

[31] TRIPATHI, D., EDLA, D. R., KUPPILI, V., ANNUSHREE BABLANI, A., AND DHARAVATH, R., *Credit scoring model based on weighted voting and cluster based feature selection*, Procedia Computer Science, 132, 22–31, (2018), https://doi.org/10.1016/j.procs.2018.05.055. (Cited on p. 91)

[32] DOSE, C., AND CINCOTTI, S., *Clustering of financial time series with application to index and enhanced index tracking portfolio*, Physica A: Statistical Mechanics and Its Applications, 355(1), 145–151, (2005), https://doi.org/10.1016/j.physa.2005.02.078. (Cited on p. 91)

[33] PATTARIN, F., PATERLINI, S., AND MINERVA, T., *Clustering financial time series: An application to mutual funds style analysis*, Computational Statistics and Data Analysis, 47(2), 353–372, (2004), https://doi.org/10.1016/j.csda.2003.11.009. (Cited on p. 91)

[34] FRICKE, D., *Trading strategies in the overnight money market: Correlations and clustering on the e-MID trading platform*, Physica A: Statistical Mechanics and Its Applications, 391(24), 6528–6542, (2012), https://doi.org/10.1016/j.physa.2012.07.045. (Cited on p. 92)

[35] PATTON, A. J., AND WELLER, B. M., *Risk price variation: The missing half of empirical asset pricing*, The Review of Financial Studies, 35(11), 5127–5184, (2022), https://doi.org/10.1093/rfs/hhac012. (Cited on pp. 92, 203)

[36] HOEL, P. G., PORT, S. C., AND STONE, C. J., *Introduction to Stochastic Processes*, Houghton Mifflin, Boston, MA, 1972. (Cited on p. 92)

[37] HIGHAM, C. F., AND HIGHAM, D. J., *Deep learning: An introduction for applied mathematicians*, SIAM Review, 61(4), 860–891, (2019), https://doi.org/10.1137/18M1165748. (Cited on pp. 20, 100)

[38] HORNIK, K., STINCHCOMBE, M., AND WHITE, H., *Multilayer feedforward networks are universal approximators*, Neural Networks, 2(5), 359–366, (1989), https://doi.org/10.1016/0893-6080(89)90020-8. (Cited on p. 102)

[39] KOLMOGOROV, A., *On the representation of continuous functions of several variables by superpositions of continuous functions of a smaller number of variables*, Proceedings of the USSR Academy of Sciences, 108, 179–182, (1956); English translation: American Mathematical Society Translations Series 2, 17, 369–373, (1961). (Cited on p. 102)

[40] ARNOLD, V., *On functions of three variables*, Proceedings of the USSR Academy of Sciences, 114, 679–681, (1957); English translation: American Mathematical Society Translations Series 2, 28, 51–54, (1963). (Cited on p. 102)

[41] CYBENKO, G., *Approximation by superpositions of a sigmoidal function*, Mathematics of Control, Signals, and Systems, 2, 303–314, (1989), https://doi.org/10.1007/BF02551274. (Cited on p. 102)

[42] BINGHAM, N., *Andrei Nikolaevich Kolmogorov*, in The Princeton Companion to Mathematics, edited by T. Gowers, J. Barrow-Green, and I. Leader, Princeton University Press, Princeton, NJ, 814–816, 2009, https://doi.org/10.1515/9781400830398.814. (Cited on p. 102)

[43] BELLMAN, R. E., *Dynamic Programming*, Princeton University Press, Princeton, NJ, 1957. (Cited on p. 120)

[44] OKSENDAL, B., *Stochastic Differential Equations: An Introduction with Applications*, 5th Edition, Springer-Verlag, New York, 2000. (Cited on p. 126)

[45] MERTON, R. C., *Optimum consumption and portfolio rules in a continuous-time model*, Journal of Economic Theory, 3(4), 373–413, (1971), https://doi.org/10.1016/0022-0531(71)90038-X. (Cited on p. 126)

[46] PHAM, H., *Continuous-Time Stochastic Control and Optimization with Financial Applications*, Springer-Verlag, Berlin, 2009. https://doi.org/10.1007/978-3-540-89500-8. (Cited on pp. 127, 168)

[47] LONGSTAFF, F., AND SCHWARTZ, E., *Valuing American options by simulation: A simple least-squares approach*, The Review of Financial Studies, 14, 113–147, (2001), https://doi.org/10.1093/rfs/14.1.113. (Cited on p. 127)

[48] TSITSIKLIS, J. N., AND VAN ROY, B., *Average cost temporal-difference learning*, Automatica, 35(11), 1799–1808, (1999), https://doi.org/10.1016/S0005-1098(99)00099-0. (Cited on p. 127)

[49] GLASSERMAN, P., *Monte Carlo Methods in Financial Engineering*, Springer, New York, 2004, https://doi.org/10.1007/978-0-387-21617-1. (Cited on p. 127)

[50] DIXON, M. F., HALPERIN, I., AND BILOKON P., *Machine Learning in Finance, From Theory to Practice*, Springer, Berlin, 2020, https://doi.org/10.1007/978-3-030-41068-1. (Cited on pp. 6, 128)

[51] GRINOLD, R. C., AND KAHN, R. N., *Active Portfolio Management: A Quantitative Approach for Producing Superior Returns and Controlling Risk*, McGraw-Hill, New York, 2000. (Cited on p. 134)

[52] CAO, L., editor, *Handbook of Artificial Intelligence and Big Data Applications in Investments*, CFA Institute Research Foundation, Charlottesville, VA, 2023, https://doi.org/10.56227/23.1.5. (Cited on pp. 134, 136, 137)

[53] KOLM, P. N., AND MACLIN, L., *Algorithmic trading, optimal execution, and dyna mic port folios*, in The Oxford Handbook of Quantitative Asset Management, edited by B. Scherer, and K. Winston, Oxford Academic, Oxford, UK, 2011, https://doi.org/10.1093/oxfordhb/9780199553433.001.0001. (Cited on p. 134)

[54] BRUNNERMEIER, M., FARHI, E., KOIJEN, R. S. J., KRISHNAMURTHY, A., LUDVIGSON, S. C., LUSTIG, H., NAGEL, S., AND PIAZZESI, M., *Review article: Perspectives on the future of asset pricing*, The Review of Financial Studies, 34(4), 2126–2160, (2021), https://doi.org/10.1093/rfs/hhaa129. (Cited on p. 135)

[55] GOLDSTEIN, I., SPATT, C. S., AND YE, M., *Big data in finance*, The Review of Financial Studies, 34, 3213–3225, (2021), https://doi.org/10.1093/rfs/hhab038. (Cited on pp. 135, 137)

[56] KAKUSHADZE, Z., AND SERUR, J. A., 151 *Trading Strategies*, Palgrave Macmillan, an imprint of Springer Nature, Cham, Switzerland, 2018. (Cited on p. 135)

[57] FAMA, E. F., *Efficient capital markets:* II, Journal of Finance, 46, 1575–1617, (1991), https://doi.org/10.1111/j.1540-6261.1991.tb04636.x. (Cited on pp. 135, 199)

[58] MARKOWITZ, H., *Portfolio selection*, The Journal of Finance, 7(1), 77–91, (1952), https://doi.org/10.2307/2975974. (Cited on pp. 135, 139)

[59] VON NEUMANN, J., AND MORGENSTERN, O., *Theory of Games and Economic Behavior*, Princeton University Press, Princeton, NJ, 1944, http://www.jstor.org/stable/j.ctt1r2gkx. (Cited on p. 152)

[60] KAHNEMAN, D., AND TVERSKY, A., *Prospect theory: An analysis of decision under risk*, Econometrica, 47(2), 263–291, (1979), https://doi.org/10.2307/1914185. (Cited on p. 153)

[61] BLACK, F., AND LITTERMAN, R., *Global portfolio optimization*, Financial Analysts Journal, 48(5), 28–43, (1992). (Cited on p. 155)

[62] COX, J. C., AND HUANG, C. F., *Optimal consumption and portfolio policies when asset prices follow a diffusion process*, Journal of Economic Theory, 49(1), 33–83, (1989), https://doi.org/10.1016/0022-0531(89)90067-7. (Cited on p. 167)

[63] KARATZAS, I., LEHOCZKY, J. P., AND SHREVE, S. E., *Optimal portfolio and consumption decisions for a "small investor" on a finite horizon*, SIAM Journal of Control and Optimization, 25, 1557–1586, (1987), https://doi.org/10.1137/0325086. (Cited on p. 167)

[64] SHREVE, S. E., *Stochastic Calculus for Finance* II*: Continuous-Time Models*, Springer, Berlin, 2004. (Cited on p. 167)

[65] ALMGREN, R., AND CHRISS, N., *Optimal execution of portfolio transactions.*, Journal of Risk, 3, 5–39, (2000). (Cited on p. 168)

[66] CARTEA A., JAIMUNGAL S., AND PENALVA J., *Algorithmic and High-Frequency Trading*, Cambridge University Press, Cambridge, UK, 2015. (Cited on p. 168)

[67] GÂRLEANU, N., AND PEDERSEN, L. H., *Dynamic portfolio choice with frictions*, Journal of Economic Theory, 165, 487–516, (2016). (Cited on pp. 168, 169)

[68] DUFFIE, D., *Dynamic Asset Pricing Theory*, Princeton University Press, Princeton, NJ, 1992. (Cited on pp. 168, 198)

[69] LUCAS, R. E., *Asset prices in an exchange economy*, Econometrica, 46(6), 1429–1445, (1978), https://doi.org/10.2307/1913837. (Cited on p. 177)

[70] SHARPE, W. F., *Capital asset prices: A theory of market equilibrium under conditions of risk*, Journal of Finance, 19, 425–442, (1964). (Cited on pp. 6, 199)

[71] CAMPBELL, J. Y., AND YOGO, M., *Efficient tests of stock return predictability*, Journal of Financial Economics, 81(1), 27–60, (2006), https://doi.org/10.1016/j.jfineco.2005.05.008. (Cited on p. 199)

[72] FAMA, E. F., AND FRENCH, K. R., *The cross-section of expected stock returns*, The Journal of Finance, 47(2), 427–465, (1992), https://doi.org/10.2307/2329112. (Cited on p. 199)

[73] FAMA, E. F., AND FRENCH, K. R., *A five-factor asset pricing model*, Journal of Financial Economics, 116(1), 1–22, (2015), https://doi.org/10.1016/j.jfineco.2014.10.010. (Cited on p. 199)

[74] BAILER, H. M., MARAVINA, T. A., AND MARTIN, D. R., *Robust betas in asset management*, in The Oxford Handbook of Quantitative Asset Management, edited by B. Scherer and K. Winston, Oxford Academic, Oxford, UK, 2012, https://doi.org/10.1093/oxfordhb/9780199553433.013.0011. (Cited on p. 200)

[75] FAMA, E. F., AND MACBETH, J. D., *Risk, return, and equilibrium: Empirical tests*, Journal of Political Economy, 81(3), 607–636, (1973). (Cited on p. 201)

[76] DEMPSTER, A. P., LAIRD, N. M., AND RUBIN, D. B., *Maximum likelihood from incomplete data via the EM algorithm*, Journal of the Royal Statistical Society, 39, 1–38, (1977), https://doi.org/10.2307/2984875. (Cited on p. 203)

[77] GIGLIO, S., LIAO, Y., AND XIU, D., *Thousands of alpha tests*, The Review of Financial Studies, 34(7), 3456–3496, (2021), https://doi.org/10.1093/rfs/hhaa111. (Cited on p. 203)

[78] TIAN, M. H., *Firm characteristics and empirical factor models: A model mining experiment*, The Review of Financial Studies, 34(12), 6087–6125 (2021), https://doi.org/10.1093/rfs/hhaa126. (Cited on p. 224)

[79] JENSEN, T. I., KELLY, B. T., MALAMUD, S., AND PEDERSEN, L. H., *Machine Learning and the Implementable Efficient Frontier*, Swiss Finance Institute Research Paper No. 22-63, 2022, https://doi.org/10.2139/ssrn.4187217. (Cited on p. 227)

[80] HANSEN, P., AND JAGANNATHAN, R., *Assessing specification errors in stochastic discount factor models*, The Journal of Finance, 52(2), 557–590, (1997). (Cited on p. 213)

[81] DAUL, S., JAISSON, T., AND NAGY, A., *Performance attribution of machine learning methods for stock returns prediction*, The Journal of Finance and Data Science, 8, 86–104, (2022), https://doi.org/10.1016/j.jfds.2022.04.002. (Cited on p. 227)

Index